Power Electronics and Power Systems

Series Editors

M. A. Pai
Alex M. Stankovic

For further volumes:
http://www.springer.com/series/6403

Vijay Vittal · Raja Ayyanar

Grid Integration and Dynamic Impact of Wind Energy

 Springer

Vijay Vittal
School of Electrical, Computer
  and Energy Engineering
Arizona State University
Tempe, AZ 85287
USA

Raja Ayyanar
School of Electrical, Computer
  and Energy Engineering
Arizona State University
Tempe, AZ 85287
USA

ISBN 978-1-4419-9322-9   ISBN 978-1-4419-9323-6   (eBook)
DOI 10.1007/978-1-4419-9323-6
Springer New York Heidelberg Dordrecht London

Library of Congress Control Number: 2012938477

Printed on acid-free paper

Springer is part of Springer Science+Business Media (www.springer.com)

Preface

Wind-based renewable energy generation has had a significant resurgence in the past decade. This is primarily due to advances in technology driven by the advent of variable speed wind turbine generators, which include doubly fed induction generator-based wind turbines as well as full converter permanent magnet synchronous machine based wind turbines. The versatility of these wind generators can be largely attributed to power electronic converters that enable variable speed operations while providing increasing grid support features. This resurgence has resulted in a significant number of wind farms being interconnected to the electric grid all around the world. As a consequence, there has been renewed interest in examining and analyzing the impact of increased penetration of wind resources on the steady-state and dynamic performance of the interconnected grid. The need to understand and carefully analyze this impact has led to concerted efforts in the development of models of different types of wind turbine generators, and the incorporation of these models into analysis tools.

Modern wind turbine generators are complex devices, and examining their impact on system behavior requires a careful study of wind turbine generators together with their associated controls. In order to achieve this objective, there is need to develop a deeper understanding of the critical components associated with wind turbine generators, including the mechanical and dynamic characteristics of wind turbines, the electrical and dynamic characteristics of generators, and the characteristics of power electronic converters and their associated controls.

The authors have been closely associated with the examination of the impact of increased penetration of wind generation on system performance with their colleagues and students. Other investigators have also made significant contributions to this field. Our objective in developing this book is to present to the reader an account of the salient aspects of the various wind turbine technologies, details of the associated power electronic converters and their controls, and a comprehensive discussion of the impact of wind turbine generators on system dynamic performance of the electric grid. We hope that this book will provide an understanding of the basics of wind turbine technology and their impact on system performance. Finally, we wish to express our gratitude to the graduate students, especially Youyuan Jiang, Siddharth Kulasekaran and Durga Gautam for their help in providing many of the figures and simulation results, and to Sunanda Vittal for her meticulous proof-reading.

Tempe, AZ, USA, February 2012   Vijay Vittal
                                                                   Raja Ayyanar

Contents

Chapter 1
Introduction

1.1 Overview of Wind Generation

Growth of wind power in the United States and around the world continues to surpass even optimistic projections of the past years with a string of record breaking years. In 2009 alone, 10 GW of new wind power capacity was added in the United States, which is 20 % higher than the record set in 2008, and represents 39 % of all new capacity added in 2009 [1]. Figure. 1.1 [1], which superimposes the actual installed wind generation with the deployment path laid out by [2] to realize the vision of 20 % wind by 2030, shows dramatically that the actual growth in the last 4 years and the projected growth in 2010–2012 far exceeded the deployment plan.

Though a slowdown was expected in 2010, resurgence is projected for 2011 and 2012. The North American Reliability Corporation (NERC) projects that 210 GW of new wind capacity is planned for construction in the next 10 years [3], which again exceeds the pace required—a total of 300 GW by 2030 for 20 % penetration. Even at present, several states have high wind penetration; for example, Iowa has 19.7 % of its total generation derived from wind resources, and nine utilities are estimated to have more than 10 % wind energy on their systems. Aggregated data on interconnection queues from various ISOs and utilities also confirm the strong interest and continued growth in wind. Figure 1.2 compares the different generation resources in 33 interconnection queues in 2009 with wind far exceeding the other resources [1].

Several countries are taking steps to develop large-scale wind markets. According to news released by the Global Wind Energy Council (GWEC), the sum of the world's total wind installations increased by 31 % to reach over 157.9 GW by the end of 2009 [4]. The increase in capacity of over 100 % from 12.1 GW in 2008 to 25.1 GW (with new capacity additions of 13 GW) by the end of 2009 made China the number one market in terms of new wind power installations. With the addition of nearly 10 GW of wind power in 2009, the United States

V. Vittal and R. Ayyanar, *Grid Integration and Dynamic Impact of Wind Energy*,
Power Electronics and Power Systems, DOI: 10.1007/978-1-4419-9323-6_1,
© Springer Science+Business Media New York 2013

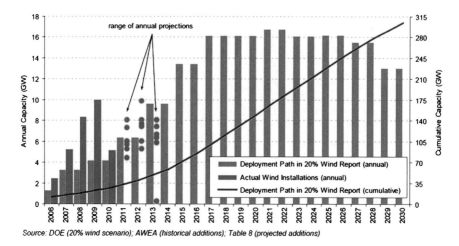

Source: DOE (20% wind scenario); AWEA (historical additions); Table 8 (projected additions)

Fig. 1.1 Actual wind installation versus deployment path required for realizing 20 % wind by 2030 [1]

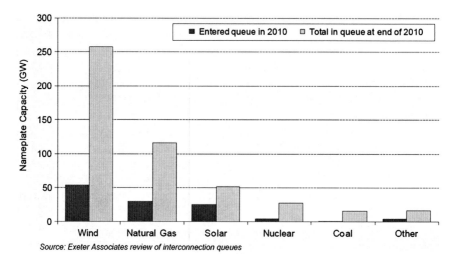

Source: Exeter Associates review of interconnection queues

Fig. 1.2 Capacity of various generation sources including wind and solar in 2010 in 33 different interconnection queues [1]

remained the leading nation in wind power in the year 2009 with 22.3 % of the world's total installed wind capacity. Following in rank were China, Germany, and Spain with installed capacities of 25.80, 25.77, and 19.14 GW, respectively. Given this scenario of significant wind penetration in the United States, also reflected in other countries around the world, it is important to examine the impact of this increased wind penetration on the performance and reliability of the electric power system.

Large-scale wind resources are interconnected to the bulk transmission system, which serves as the primary instrument to transfer the energy generated from the wind resources to the load centers. Most of the available wind generator technologies generate electricity asynchronously. In contrast to conventional synchronous generators, this implies that the position of the rotor flux vector is not dependent on the physical position of the rotor, and the synchronizing torque angle characteristic does not exist [5]. This salient feature of wind turbine generators coupled with the variability of wind resources introduces complexities and factors that need to be carefully analyzed to understand and evaluate the impact of increased wind penetration on power system performance.

Wind turbines can be classified based on the technology used as either fixed speed or variable speed. A fixed speed wind turbine is directly connected to the grid system. A variable speed turbine, however, is interfaced to the grid using power electronic equipment. There are several advantages to using a variable speed turbine, including lower stresses on the mechanical structures, reduction of acoustic noise, and the ability to control both active and reactive power output. The current boom in wind energy can be attributed to a large extent to the availability of large wind turbines typically rated in the 1.5–5 MW range. Modern large wind turbines are all variable speed machines. They typically incorporate pitch control and include either a doubly fed induction generator (DFIG) or a permanent magnet synchronous generator. Presently, the DFIG technology is most widely adopted among wind turbine manufacturers for larger wind turbines and is the predominant technology in most wind farms developed since 2005. The DFIG technology has a relatively low capital cost. However, its disadvantages include suboptimal power quality, reduced reliability resulting from the gearboxes, and the need for unique gearbox and generator requirements for systems operating at 50 Hz versus 60 Hz.

1.2 Wind Turbine Generator Technologies

Four wind turbine generator technologies are commercially prevalent and available for utility size applications. They include:

1. Fixed speed wind generators with squirrel cage induction generators—Type 1
2. Wind generators with wound rotor induction generators and limited speed variation through an external resistor—Type 2
3. Doubly fed induction generators with variable speed—Type 3
4. Permanent magnet synchronous machine or an induction machine (cage or wound rotor) with a full converter and variable speed range—Type 4

A brief description of each of these wind turbine generator technologies is now provided. DFIG-based Type 3 machines are currently the most predominantly used wind turbine generators. The model and controls associated with these types of generators will be explained in Sect. 1.3.

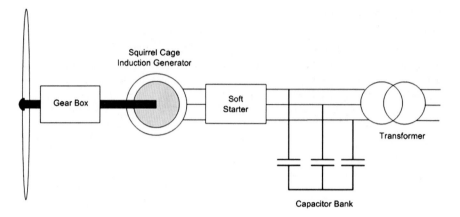

Fig. 1.3 Schematic of Type 1 wind turbine generator

1.2.1 Type 1 Wind Turbine Generators

The schematic of a Type 1 wind turbine generator is shown in Fig. 1.3. In these types of machines, the induction generator is directly interconnected to the electrical grid. The gear box and the number of pole pairs of the induction generator determine the fixed speed of the wind turbine generator.

The Type 1 machines are the least expensive of the wind turbines. Their construction does not involve any slip rings and hence the machines are rugged and reliable. These machines, however, have suboptimal energy extraction because of the fixed speed operation. In addition, capacitor banks are required to provide reactive power support. In order to overcome the disadvantage of the suboptimal energy extraction, a variation in the architecture has been implemented by Danish manufacturers [6]. The improvement in energy extraction is accomplished by utilizing two generators of different ratings and pole pairs or using a single generator with two separate windings having different ratings and pole pairs. This arrangement results in increased energy extraction and also reduces the magnetizing losses at low wind speeds.

1.2.2 Type 2 Wind Turbine Generators

Type 2 wind turbine generators typically include a wound rotor induction generator with a variable external resistance connected in series with the rotor winding. A limited range of speed variation is achieved through the external resistor [7]. The torque-speed curve of the machine can be manipulated by varying the external resistance. The rotor resistor is mounted on the generator shaft and controlled using optical signals. Hence, the need for slip rings is avoided. The external

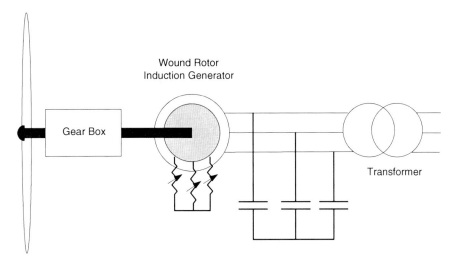

Fig. 1.4 Schematic of Type 2 wind turbine generator

resistor is pulsed using a DC chopper circuit with a variable duty cycle. This results in a variable resistance with no moving parts. This arrangement provides a limited variation of generator speed by approximately 10 % in order to provide a constant power output with change in wind speeds. The wind generator always operates at a variable slip above synchronous speed. This arrangement is not as flexible as a doubly fed induction generator but results in significantly less cost. The machine still lacks the ability to control reactive power consumption independent of the real power and an external capacitor bank has to be provided. In addition, the inclusion of the external resistance results in higher rotor losses, which increases with slip and limits the range of speed variation. A schematic of the Type 2 wind turbine generators is shown in Fig. 1.4.

1.2.3 Type 3 Wind Turbine Generators

This category of wind turbine generators is a variable speed machine that includes a wind turbine with a double fed induction generator. Wound rotor induction machines are utilized, and the stator is directly connected to the grid. The rotor winding is connected using slip rings to a machine side converter. The machine side converter is coupled through a DC bus capacitor to the grid side converter which is connected to the grid via a transformer. The mechanical speed of the machine can be controlled by operating the rotor circuit at a variable frequency. The net power output from the machine in this design is the sum of the power from the machine's stator and that from the rotor (through the converter) into the grid. Power is injected from the rotor, through the converter, into the grid when the

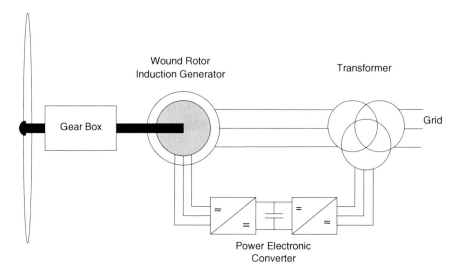

Fig. 1.5 Schematic of Type 3 wind turbine generator

machine operates at supersynchronous speed. When the machine operates at subsynchronous speeds, however, real power is absorbed from the grid through the converter by the rotor. At exactly synchronous speed, no significant net power exchange occurs between the rotor and the grid.

Most Type 3 designs have the capability to provide reactive power support to the grid through the machine's stator. This capability is achieved by changing the d-axis excitation on the rotor. A vector control approach is utilized to split the rotor current into a d-axis (flux producing) component and q-axis (torque producing) component. Each of these components is then controlled separately. The power factor of machine is regulated by controlling the d-axis component and the electrical torque of the machine is held fixed by controlling the q-axis component. An alternative approach to provide reactive power is through the use of a four-quadrant voltage source converter design in the grid side converter, which would then act as STATCOM and supply or absorb reactive power as needed. This device could function irrespective of whether the actual wind turbine generator was operational or disconnected from the system. The provision of this additional feature would result in additional costs. A schematic of a Type 3 wind turbine generator is shown in Fig. 1.5.

Type 3 wind turbine generators are commonly referred to as DFIG wind turbine generators. They are currently the most commonly adopted wind turbine generators in wind farms around the world. The detailed modeling of these generators will be discussed in greater detail later in this chapter.

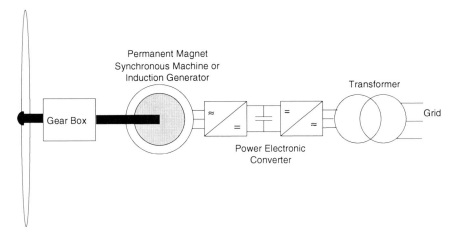

Fig. 1.6 Schematic of Type 4 wind turbine generator

1.2.4 Type 4 Wind Turbine Generators

This class of wind turbine generators is also a variable speed machine equipped with a fully rated converter that interconnects the stator of the machine to the grid. The generator could either be a squirrel cage or wound rotor induction generator or a permanent magnet synchronous generator. These types of generators have a wide speed range and are capable of maximum power extraction. In addition, this class of machines has independent active and reactive power control. Type 4 machines in general are more expensive because they include two power converters processing fully rated power. Figure 1.6 depicts a schematic of the Type 4 wind turbine generator. This type of wind turbine generator is slowly gaining wide acceptance and could become a standard fixture in the future wind farms since it allows for operation over a wide range of wind speeds and gearless design. These could result in significant advantages, which could overcome the increased cost.

1.3 Detailed Representation and Modeling of Type 3 Wind Turbine Generators

Type 3 wind turbine generators are currently the most widely used technology in the field and have found acceptance internationally. Keeping this fact in mind, significant details regarding the functioning, detailed mathematical modeling, control representation, and modeling for electromechanical time domain simulation will be provided.

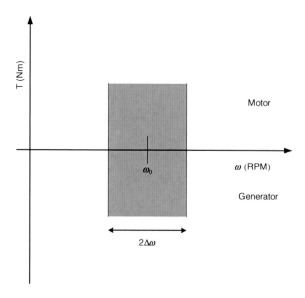

Fig. 1.7 Torque-speed characteristic of a DFIG

Type 3 wind turbine generators using a doubly fed induction generator (DFIG) provide a viable solution for variable speed configuration requiring a limited variable speed range of about ±30 % of synchronous speed. The primary limitation with regard to this speed variation range is that the power electronic converter is limited to handle a fraction (20–30 %) of the total power. This feature also results in reduced losses in the power electronic converter in comparison to Type 4 units in which the converter handles the total power of the unit. As a result, the cost of the converter is lower. As depicted in Fig. 1.5, in Type 3 units the stator circuit of the DFIG is directly connected to the grid. The rotor circuit is connected by means of slip rings to a back-to-back converter consisting of two converters, a machine side converter and a grid side converter. A DC-link capacitor is placed for energy storage between the two converters. This capacitor limits the voltage variations in the DC-link. The torque or the speed of the DFIG and also the power factor at the stator terminals can be controlled using the machine side converter while the grid side converter controls the DC-link voltage. The torque-speed characteristic of the DFIG wind turbine generator is shown in Fig. 1.7 [8].

As depicted in Fig. 1.7 the DFIG machine can operate either as a motor or as a generator at both subsynchronous and supersynchronous speeds. For the cases of a DFIG operating as a generator at both subsynchronous and supersynchronous speeds the configurations of the real power flows in the various circuits of the machine are depicted in Figs. 1.8 and 1.9.

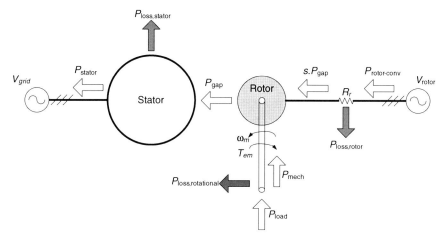

Fig. 1.8 DFIG power exchange at subsynchronous speeds

1.3.1 DFIG as a Generator at Subsynchronous Speeds

Figure 1.8 depicts the real power exchanges and the associated losses when the DFIG acts as a motor at subsynchronous speed. The slip s of the machine is positive and is given by

$$s = \frac{n_s - n_r}{n_s} \qquad (1.1)$$

where

n_r	= rotor speed in revolutions per minute (rpm)
$n_s = 120f/p$	= synchronous speed (rpm)
f	= electrical frequency of stator current
p	= number of poles

$$n_r = (1 - s)n_s = \left(1 - \frac{f_r}{f}\right)n_s \qquad (1.2)$$

where
f_r = frequency of rotor current

The electromagnetic torque T_{em} is given by

$$T_{em} = \frac{P_{gap}}{\omega_{syn}} \qquad (1.3)$$

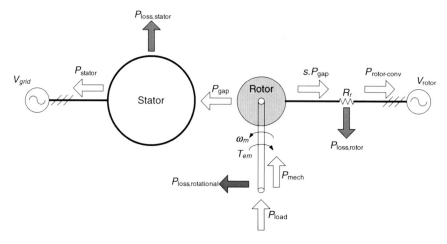

Fig. 1.9 DFIG power exchange at supersynchronous speeds

where

P_{gap} = power transferred through the air gap

$\omega_{\text{syn}} = 2\pi f$ = synchronous frequency in radians/s

The mechanical power P_{mech} is given by

$$P_{\text{mech}} = (1 - s)P_{\text{gap}} = \left(1 - \frac{f_r}{f}\right)P_{\text{gap}} \tag{1.4}$$

1.3.2 DFIG as a Generator at Supersynchronous Speeds

The real power exchanges when a DFIG operates as a generator at supersynchronous speeds is shown in Fig. 1.9.

In this case the slip s given by (1.1) is negative. The rotor speed n_r is given by

$$n_r = (1 - s)n_s = \left(1 + \frac{f_r}{f}\right)n_s \tag{1.5}$$

and the mechanical power P_{mech} is given by

$$P_{\text{mech}} = (1 - s)P_{\text{gap}} = \left(1 + \frac{f_r}{f}\right)P_{\text{gap}} \tag{1.6}$$

1.3.3 Wind Power Model

The mechanical power extracted from the wind by a wind turbine is a complex function of the wind speed, blade pitch angle, and shaft speed. The algebraic equation shown below characterizes the power extracted.

$$P_m = \frac{1}{2} \rho v_w^3 \pi r^2 C_p(\lambda) \tag{1.7}$$

where
P_m = the power extracted from the wind, in watts
ρ = the air density, in kg/m^3
r = the radius swept by the rotor blades, in m
v_w = wind speed, in m/s
C_p = the performance coefficient
λ = the tip speed ratio, i.e., the ratio of turbine blade speed to that of the wind

$$\lambda = \frac{\omega_t r}{v_w} \tag{1.8}$$

where
ω_t = mechanical rotor speed radians/s

   From Eq. (1.7) it is noted that the air density, the wind speed, and the radius swept by the blades are not quantities that can be controlled. Hence, the only parameter that can be controlled in order to maximize the energy output from the wind is the performance coefficient C_p, which has a theoretical maximum governed by Betz's law of 0.593.
   The performance coefficient C_p for a given blade pitch angle and rotation speed is nonlinearly related to the wind speed. C_p typically peaks at a given turbine tip speed to wind speed ratio and then drops off again to zero at higher tip speed ratios. The C_p characteristic is manufacturer-specific and varies for turbines provided by different manufacturers. A typical plot of C_p versus the tip speed ratio λ is shown in Fig. 1.10, where the change of C_p as pitch angle is adjusted is also shown
   At low to medium wind speeds the pitch angle is controlled to allow the wind turbine to operate at its optimum condition. In high wind speed regions, the pitch angle is adjusted to spill some of the aerodynamic energy. A wind turbine is typically designed to extract the maximum amount of wind energy possible at wind speeds in the range 10–15 m/s. At speeds beyond 15 m/s most modern turbines spill some of the energy and cut out completely at wind speeds in excess of 20–25 m/s. In Type 3 machines this is achieved by utilizing dynamic blade pitch-control. As a result of this control turbine speed is adjusted based on the wind speed to maximize the electric power output. The operation at the maximum power point is realized over a wide power range. A typical set of output power-

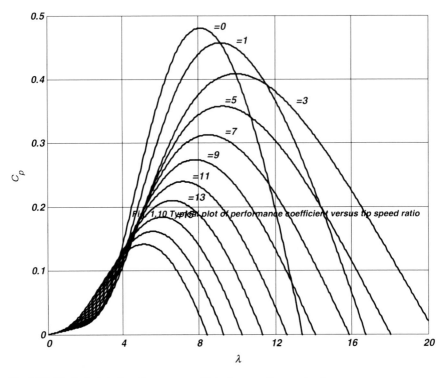

Fig. 1.10 Typical plot of performance coefficient versus tip speed ratio

speed curves as a function of turbine speed and wind speed is shown in Fig. 1.11. In this figure, the electric power output and turbine speed are normalized using their respective rated quantities.

From Eq. (1.8) it can be observed that the tip speed ratio for a given wind turbine speed varies over a wide range depending on the wind speed. However, from Eq. (1.7) it can be seen that the power production from the wind turbine can be maximized if the turbine is operated at maximum C_p. In order to achieve this objective, the rotor speed should be adjusted to follow the change in wind speed. The variable speed DFIG technology is capable of doing this. The rotor speed can be controlled by controlling the difference between the electrical output power and the power extracted from the wind. With dynamic pitch control, the power extracted from the wind can be controlled, and with the use of a power electronic converter, the electrical power output can also be controlled. As a result, the rotor speed can be controlled. A detailed development of the dynamic pitch control is presented in [9].

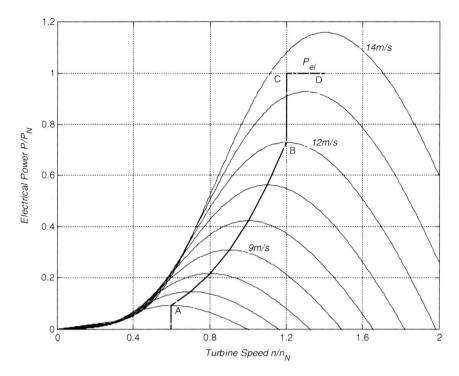

Fig. 1.11 Plot of electrical output power as a function of wind turbine speed

1.3.4 Mechanical Drive Train Model

The drive train system of a wind turbine consists of the rotating masses, hub, gearbox, connecting shafts, and the generator inertia. The detailed model is a six-mass model [10]. For power system studies [11], a two-mass model of the drive train shown in Fig. 1.12 is needed. The two-mass model captures the effects of the wind turbine and the generator.

The equations of motion are obtained applying Newton's law of motion for each mass and shaft as follows:

$$\begin{cases} T_e - D_{tg}(\omega_e - \omega_t) - K_{tg}(\theta_m - \theta_t) = 2H_g\dot{\omega}_e \\ D_{tg}(\omega_e - \omega_t) + K_{tg}(\theta_e - \theta_t) - T_t = 2H_t\dot{\omega}_t \end{cases} \tag{1.9}$$

where ω_e and ω_t [both in per unit] are the generator and turbine speeds, respectively; $(\theta_e - \theta_t)$ is the shaft twist angle; H_g and H_t are the generator and turbine inertias, respectively; K_{tg} [per unit torque/rad] and D_{tg} [per unit torque/per unit speed] are the shaft stiffness and damping coefficients, respectively; and T_t and T_e are the turbine torque and electric torque, respectively.

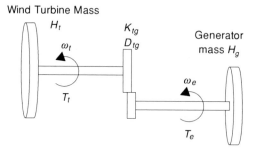

Fig. 1.12 Two-mass model of drive train

1.3.5 Modeling of Doubly Fed Induction Generator

A commonly used mathematical model of the induction generator in the $d-q$ reference frame is developed. This model is also derived in detail in [12]. The schematic of the equivalent circuit for the model is shown in Fig. 1.13. Here, v_{sq}, v_{sd} are stator voltages and v_{rq}, v_{rd} are rotor voltages referred to the stator side, λ_{sd}, λ_{sq} are the stator fluxes and λ_{rd}, λ_{rq} are rotor fluxes referred to the stator side, i_{sd}, i_{sq} are stator currents and i_{rd}, i_{rq} are rotor currents referred to the stator side, R_s is the stator resistance, R_r is the rotor resistance referred to the stator side, L_{ls}, L_{lr} are the stator and stator-referred rotor leakage inductances, respectively, L_m is the mutual inductance, ω is the reference frame speed, and ω_r is the rotor electrical frequency. Here, the motor convention is used, which means that a current entering the machine is positive whereas a current leaving the machine is negative, and the $d-q$ reference frame is defined as the q-axis leading the d-axis by 90°.

1.3.5.1 Voltage and Flux Equations

Equations (1.10) and (1.11) are voltage and flux equations of the DFIG expressed in motor convention, with all the quantities on the rotor side referred to stator side.

$$\begin{aligned}
v_{sq} &= R_s i_{sq} + \omega \lambda_{sd} + \dot{\lambda}_{sq} \\
v_{sd} &= R_s i_{sd} - \omega \lambda_{sq} + \dot{\lambda}_{sd} \\
v_{rq} &= R_r i_{rq} + (\omega - \omega_r)\lambda_{rd} + \dot{\lambda}_{rq} \\
v_{rd} &= R_r i_{rd} - (\omega - \omega_r)\lambda_{rq} + \dot{\lambda}_{rd}
\end{aligned} \qquad (1.10)$$

$$\begin{aligned}
\lambda_{sq} &= L_s i_{sq} + L_m i_{rq} \\
\lambda_{sd} &= L_s i_{sd} + L_m i_{rd} \\
\lambda_{rq} &= L_r i_{rq} + L_m i_{sq} \\
\lambda_{rd} &= L_r i_{rd} + L_m i_{sd}
\end{aligned} \qquad (1.11)$$

where $L_s = L_{ls} + L_m$, $L_r = L_{lr} + L_m$.

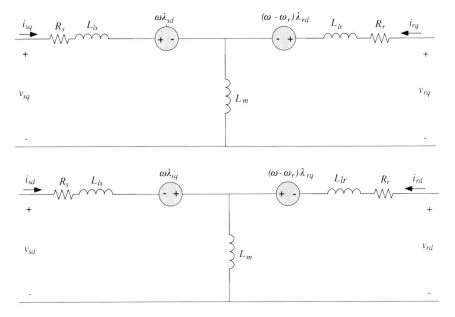

Fig. 1.13 Equivalent model of doubly fed induction generator

1.3.5.2 Torque Equation

The electromagnetic torque of the induction machine is then given by

$$T_e = \frac{3}{2}\frac{p}{2}L_m\left(i_{sq}i_{rd} - i_{sd}i_{rq}\right) \tag{1.12}$$

where p is the number of poles of the machine.

1.3.5.3 Stator-to-Rotor Turns Ratio

In both Eqs.(1.11) and (1.12), all the rotor quantities are referred to stator side. So stator-to-rotor turns ratio should be considered in the process of DFIG modeling. The turns ratio is expressed as $n_s/n_r = v_{nom}/v_{nom\_r}$, in which v_{nom} is the rated RMS value of stator side line-to-line voltage, while v_{nom_r} is the rated RMS value of rotor side line-to-line voltage. So the implementation of the DFIG model is realized by multiplying the rotor-side-converter voltage by n_s/n_r to refer the voltage to stator side, and multiplying the stator-referred current in the DFIG model by n_s/n_r to convert the stator-referred rotor currents back to rotor currents.

   The DFIG model is realized here with mainly the voltage equations, flux equations, and the electromagnetic torque expression, while the swing equation is integrated into the two-mass model of the wind turbine.

Table 1.1 Per unit bases of the machine

Quantity	Expression
$v_{s\text{ base}}$	peak value of rated phase voltage, V
$i_{s\text{ base}}$	peak value of rated phase current, A
f_{base}	rated frequency, Hz
ω_{base}	$2\pi f_{\text{base}}$, elec. rad/s
$\omega_{m\text{ base}}$	$\omega_{\text{base}}(2/p)$, mech. rad/s
$Z_{s\text{ base}}$	$v_{s\text{ base}}/i_{s\text{ base}}$, Ω
$L_{s\text{ base}}$	$v_{s\text{ base}}/(i_{s\text{ base}}\omega_{\text{base}})$, H
$\Psi_{s\text{ base}}$	$v_{s\text{ base}}/\omega_{\text{base}}$, Wb·turns
3-phase VA$_{\text{base}}$	$3/2(v_{s\text{ base}}\, i_{s\text{ base}})$, VA
Torque base	$3/2(p/2)\Psi_{s\text{ base}}\, i_{s\text{ base}}$, N·m

1.3.5.4 Per Unit System for the Induction Machine

The base values for the induction machine are shown in Table 1.1. Utilizing these base values, the voltage, flux, and the torque equations can all be converted into common system base as is normally done for simulating large multi-machine power systems.

Based on the per unit system, the voltage equation, flux equation, and the torque equation of the induction machine can be expressed in per unit system as:

$$v_{sq} = R_s i_{sq} + \omega\lambda_{sd} + \frac{1}{\omega_{\text{base}}}\dot{\lambda}_{sq}$$
$$v_{sd} = R_s i_{sd} - \omega\lambda_{sq} + \frac{1}{\omega_{\text{base}}}\dot{\lambda}_{sd}$$
$$v_{rq} = R_r i_{rq} + (\omega - \omega_r)\lambda_{rd} + \frac{1}{\omega_{\text{base}}}\dot{\lambda}_{rq} \tag{1.13}$$
$$v_{rd} = R_r i_{rd} - (\omega - \omega_r)\lambda_{rq} + \frac{1}{\omega_{\text{base}}}\dot{\lambda}_{rd}$$

$$\lambda_{sq} = L_s i_{sq} + L_m i_{rq}$$
$$\lambda_{sd} = L_s i_{sd} + L_m i_{rd}$$
$$\lambda_{rq} = L_r i_{rq} + L_m i_{sq} \tag{1.14}$$
$$\lambda_{rd} = L_r i_{rd} + L_m i_{sd}$$

$$T_e = L_m(i_{sq}i_{rd} - i_{sd}i_{rq}) \tag{1.15}$$

Here, all the quantities are in per unit on the machine base except ω_{base}.

1.4 Controls for Type 3 Wind Turbines

In Type 3 wind turbine generators, there are three main controllers that provide controls for frequency/active power, voltage/reactive power, and pitch angle/ mechanical power. In order to provide a more meaningful description of these controllers and their physical impact on the wind turbine system, a slightly more detailed description of the power electronic converters and the controller mechanisms and functions is needed. This will be provided in Chaps. 2 and 3. The controller details will be developed in Chap. 4.

References

1. Wiser R, Bolinger M (2011) 2010 Wind technologies market report, DOE EERE Report, Available via http://www1.eere.energy.gov/wind/pdfs/51783.pdf. June 2011
2. 20% Wind Energy by 2030: Increasing Wind Energy's Contribution to U.S. Electricity Supply, DOE Report, DOE/GO-102008-2567, July 2008
3. 2009 Scenario Reliability Assessment 2009-2001, North American Electric Reliability Corporation (NERC) Report, Oct 2009
4. Global Wind Energy Council (GWEC) (2009) Global wind 2009 report, Available via http://www.gwec.net/fileadmin/documents/Publications/Global_Wind_2007_report/GWEC_Global_Wind_2009_Report_LOWRES_15th.%20Apr.pdf.
5. Hughes FM, Lara OA, Jenkins N, Strbac G (2005) Control of DFIG—based wind generation for power network support. IEEE Trans Power Sys 20(4):1958–1966 Nov 2005
6. Hansen LH, Helle L, Blaabjerg F, Ritchie E, Munk-Nielsen S, Bindner H, Sørensen P, Bak-Jensen B (2001) Conceptual survey of generators and power electronics for wind turbines, Risø National Laboratory, Roskilde, Rep. Risø-R-1205(EN), ISBN 87-550-2743-8, Dec 2001
7. Burnham DJ, Santoso S Muljadi E (2009) Variable rotor-resistance control of wind turbine generators. IEEE PES GM. doi: 10.1109/PES.2009.5275637
8. Leonhard W (1996) Control of electrical drives, 2nd edn. Springer, Berlin
9. Muljadi E, Butterfield CP (2000) Pitch-controlled variable-speed wind turbine generation, NREL/CP-500-27143, Available via http://www.nrel.gov/docs/fy00osti/27143.pdf
10. Muyeen SM, Tamura J, Murata T (2009) Stability augmentation of a grid-connected wind farm. Springer, Berlin
11. Cigré Report 328 (2007) Modeling and dynamic behavior of wind generation as it relates to power system control and dynamic performance, Working Group C4.601, Aug 2007
12. Krause PC, Wasynczuk O, Sudhoff SD (2002) Analysis of electric machinery and drive systems, 2nd edn. Wiley, New York

Chapter 2
Power Electronic Concepts

The distinguishing feature of wind and photovoltaic generators compared with conventional generators is that they are controlled and interfaced to the grid through power electronic converters. The characteristics and performance of these generators, both static and dynamic, are to a large extent determined by the design of the power electronic converters. Hence, it is important to have a good understanding of the operating principles, characteristics, and basic design of these converters in the study of integration of wind and PV systems and their impact on power systems. The power converters for wind energy systems are based on voltage source converter (VSC) topologies employing pulse width modulation (PWM) methods at a relatively high frequency, usually in the order of a few kHz for the utility-scale applications considered here. Therefore, the focus of this chapter is on understanding the basic principles of operation of typical voltage source converters used in grid interface applications.

2.1 Components of a Power Electronic Converter System

A VSC-based power electronic system is a closed-loop, feedback controlled power processing system, which achieves various control and optimization objectives with high power conversion efficiency through the process of high-frequency switching and pulse-width modulation. The switching frequency can range from a few hundred Hz in high-power applications to a few MHz in low-power DC–DC converter applications. Applications involving grid integration of wind and solar resources above a few hundred kW power level typically use switching frequencies in the range of a few kHz.

The main components of a generic VSC-based power electronic converter system are illustrated in Fig. 2.1. A single power converter can possibly interface several external power systems as indicated, and the power flow among these

V. Vittal and R. Ayyanar, *Grid Integration and Dynamic Impact of Wind Energy*, Power Electronics and Power Systems, DOI: 10.1007/978-1-4419-9323-6_2, © Springer Science+Business Media New York 2013

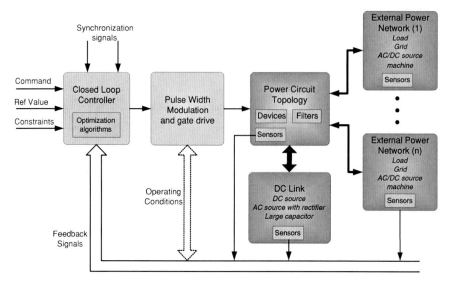

Fig. 2.1 Main components of a generic voltage source converter-based power electronic converter system

systems can be bidirectional depending on the application and operating conditions. As an example, for a DFIG wind generator, the external power systems are the three-phase power grid and the rotor windings of the induction generator. The main components shown in Fig. 2.1 are discussed briefly below.

2.1.1 Feedback Controller

The feedback controller is responsible for achieving the required control and optimization objectives specific to the given application. For renewable energy grid integration applications, the control objectives include maximizing the power captured from the renewable resource under all operating conditions, injecting power at the desired power factor, grid support functions including reactive power support, voltage regulation, and fault ride through. The design of the controller completely determines the dynamic performance and stability of the power conversion system.

The inputs to the controller include command or reference signals that are preprogrammed or derived from a higher level system controller, and various feedback signals (voltage, current, speed, power, power factor) from the converter circuits, DC link, and external power systems. The controller uses the error between the command signals and the corresponding measured values to generate suitable control signals for the pulse-width modulator stage. The controller block includes synchronization routines (phase-locked loop) and maximum power point tracking (MPPT) and other optimization algorithms. The overall control is implemented using several parallel and cascaded control loops. State-of-the-art power converters invariably use digital controllers, such as microcontrollers, digital signal processors (DSP), and field programmable grid arrays (FPGA) for the control implementation. The controller is designed, in general, based on the linearized, average model of the switching converter, filters, and external systems and loads. Controller design specific to wind energy applications is discussed in detail in Chap. 4.

2.1.2 Pulse-Width Modulator

The switch mode power electronic converters employ high-frequency switching, and use the process of pulse-width modulation (PWM) to control the average value (over a switching period) of the appropriate switching output voltages in order to realize various control objectives. We focus on constant switching frequency methods since most of the converters considered here operate at fixed switching frequency. The concept of PWM, in general, refers to controlling the ratio of ON interval of a switch pole to the total switching period.

The function of the pulse-width modulator in Fig. 2.1 is to generate the PWM signals for the gate drive circuitry of each of the switches based on the control signal provided by the controller block. An example of this process is illustrated in Fig. 2.2 for a three-phase converter. The carrier-based approach, where a modulating signal (controller output in Fig. 2.2) is compared with a triangular carrier wave to generate PWM pulses, is a popular PWM method. The frequency of the carrier wave, f_c, needs to be significantly higher than the frequency of the modulating signal, f_m. This process is explained in detail in Sect. 2.2.2. For three-phase converters, an alternate method called space vector modulation (SVM) is also widely used.

The design of the pulse-width modulator has a significant impact on the steady-state performance of the converter. PWM methods and the value of switching frequency determine the high-frequency ripple content in the output voltage and current waveforms, and therefore, impact the size of filters needed to meet the power quality standards. The PWM design also directly affects the switching losses, and therefore, the power conversion efficiency of the system.

In this book, detailed discussions are limited to two-level voltage source converters. It may be noted that multilevel converters with more number of switches and improved power quality are also beginning to be used in renewable energy integration applications. The PWM concepts, both carrier-based and space-vector-based, described for two-level converters can also be extended to multilevel converters.

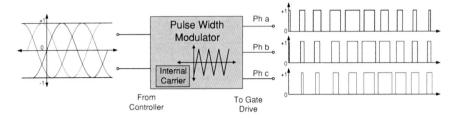

Fig. 2.2 Concept of pulse-width modulation for a three-phase system

2.1.3 Power Converter Circuit Topology

The converter topology mainly refers to the way the various power semiconductor switches and different passive components, including inductors, capacitors, and transformers (if any) are configured and operated to realize various functionalities. There are hundreds of practical converter topologies for different DC–DC, AC–DC, DC–AC, and AC–AC conversion applications. Here, we focus on the two-level bridge type topologies that include the half-bridge, full-bridge, and three-phase bridge converters as shown in Fig. 2.3, since these are the most relevant for the applications considered in this book. These topologies also enable easy understanding of the basic concepts of switch mode power conversion. Moreover, the power pole seen in each of these topologies is also a common building block for most converter topologies; hence, the analysis of a power pole here can be used to analyze many other topologies as well. The power pole as shown in Fig. 2.3a consists of two semiconductor switches that together realize a single-pole-double-through (SPDT) switch. The individual switches are turned on and off at the appropriate switching instants by an elaborate PWM and gate drive circuitry. In addition to the power semiconductor devices, the circuits consist of large capacitors to reduce pulsations in the DC link voltage, and L, LC, or LCL filters.

   For the purposes of analysis and understanding of power converter operation and its impact, we will assume the power devices and passive components to be ideal. However, practical devices have several non-idealities, including switching losses, conduction losses, finite on-state voltage drops, and switching transition and delay times, which impact converter efficiency and performance. In addition, the finite voltage, current and power (loss) ratings of the switches, and therefore the converter, should be considered in the design and analysis of the overall system.

   The principles of operation and model of a power pole are discussed in Sect. 2.2. The mechanism by which various output quantities such as output current, reactive power, and power factor are controlled using power poles is illustrated through several examples in the later sections. Since the power pole is the building block for many widely used topologies, this analysis can be readily used in analyzing most other converters.

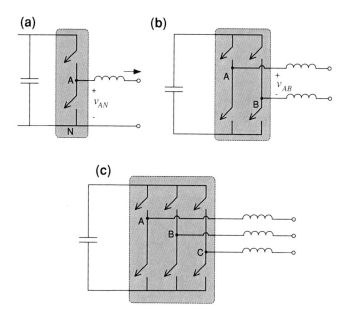

Fig. 2.3 Common power converter topologies using one or more power poles

2.1.4 DC Link and Interface with External Power Systems and Loads

VSCs operate by synthesizing controlled, low frequency voltages, in an average sense, at the output of power poles, by switching a DC voltage available across the two ends of the power poles. This DC voltage is referred to here as the DC link or DC bus, and can be derived in several ways. It can be a DC voltage source such as battery or output of an AC–DC rectifier with an AC source. However, it does not necessarily have to be a source of active power; in AC–DC rectifiers, for example, the DC link can be a load port. The average power exchanged at the DC link in general can be positive, negative, or zero, i.e., the DC link can be a power source, a power sink, or support only reactive power exchange with zero average value. In many cases the DC link can be maintained constant by large capacitors and with the power pole connected to it supplying net active power to compensate for small losses in the DC link.

The power pole itself and the converters, comprised of several power poles, can support bidirectional power flow. The instantaneous and average output voltage of an individual power pole is unipolar, but the current through the pole can be bidirectional, resulting in bidirectional power flow capability. Converters with multiple power poles can support both bipolar voltages and bidirectional currents, resulting in full four-quadrant capability. The power converter can interface two or more external power systems. The term external power systems can refer to the electric grid, or to other DC or AC voltage sources that can both source and sink power depending on operating conditions, or to the loads. This general term is used to accommodate the

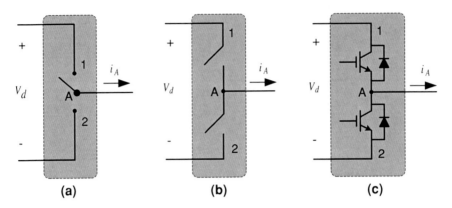

Fig. 2.4 Configuration of a power pole. **a** SPDT version, **b** double SPST implementation, and **c** electronic implementation

fact that there is no fixed source-end and load-end, with the roles reversing dynamically depending on the operating conditions at a given time. For example, in a DFIG system the power flow into the rotor winding through the rotor side PWM converter is positive during low wind speeds (subsynchronous mode) and the power flow is from the DC link to the rotor terminals. At high wind speeds (supersynchronous mode) the power flow is from the rotor terminals to the DC link.

2.2 Analysis of a Power Pole

The power pole is essentially a single-pole, double-throw (SPDT) switch. Since, there is no single electronic equivalent of an SPDT, it is implemented using two single-pole, single-throw (SPST) (simple on/off switches) switches. Each of the SPSTs is implemented electronically using a combination of a controlled switch (for example, an IGBT) and a diode. Figure 2.4 shows the SPDT version, double SPST version, and the actual electronic implementation of a power pole. The combination of two controlled switches and two diodes as shown in Fig. 2.4c supports bidirectional pole current, i_A.

There are two main requirements for the power pole of a VSC—the voltage across the two positions, indicated as V_d in Fig. 2.4 should be a non-pulsating, smooth DC voltage, and the current through the pole, indicated as i_A in Fig. 2.4 should be a non-switching current (i.e., predominantly a DC current or a low-frequency AC current) [1, 2]. The two switches of the power pole are switched at high frequency with a complementary switching pattern, i.e., when the top switch is ON, the bottom switch is OFF, and vice versa. The switching signal used to control the power pole A is denoted as $q_A(t)$. This signal directly drives the gate drive of the top switch, while its inverse drives the gate drive of the bottom switch. In practice, there is a small dead time used after a switch is turned OFF and before the next switch of the power pole is turned ON, in order to avoid the possibility of both the switches conducting during transitions, thereby resulting in large shoot-through currents. However, for the analysis presented here this dead time is ignored.

2.2.1 Switching Signal and Duty Ratio

The switching signal $q_A(t)$ is defined in Eq. (2.1) below with reference to Fig. 2.4b.

$q_A(t) = 1 \Rightarrow$ top switch ON, pole connected to position $1 \Rightarrow v_{AN}(t) = V_d$

$q_A(t) = 0 \Rightarrow$ bottom switch ON, pole connected to position $2 \Rightarrow v_{AN}(t) = 0$

$$(2.1)$$

Figure 2.5 shows a sample plot of $q_A(t)$ and the resulting $v_{AN}(t)$.

It may be noted that from a control perspective and for system level analysis, both steady state and dynamic, we are mostly interested in the average values of different converter variables. The cycle-by-cycle average (CCA) of any variable $x(t)$ is defined as its average value over exactly one switching period T_s as given in Eq. (2.2). The CCA value of a variable is denoted by a bar (-) on top as seen in Eq. (2.2)

$$\bar{x}(t) = \frac{1}{T_s} \int_{t-T_s}^{t} x(\tau) \, d\tau \qquad (2.2)$$

The duty ratio of the power pole in a given switching period refers to the ratio of the time for which $q = 1$ to the total period T_s. Referring to Fig. 2.5, the duty ratio can be defined as a continuous variable $d(t)$ as given in Eq. (2.3), where $T_{ON}(t)$ as a continuous variable denotes the duration for which $q = 1$ during the $t - T_s$ interval.

$$d(t) = \frac{T_{ON}(t)}{T_s} \qquad (2.3)$$

The CCA value of the switching signal $q_A(t)$ is derived in Eq. (2.4) to be equal to the duty ratio.

$$\bar{q}_A(t) = \frac{1}{T_s} \int_{t-T_s}^{T_s} q_A(\tau) \, d\tau = \frac{1}{T_s} \left[\int_{t-T_s}^{t-T_s+T_{ON}} 1 \, d\tau + \int_{t-T_s+T_{ON}}^{t} 0 \, d\tau \right] \qquad (2.4)$$

$$= \frac{T_{ON}(t)}{T_s} = d(t)$$

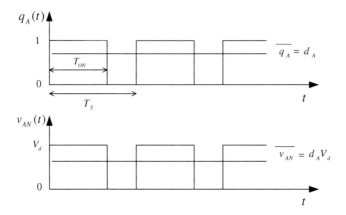

Fig. 2.5 Switching signal and the output voltage of a power pole

From Eq. (2.1), or by observing the waveforms shown in Fig. 2.5, the instantaneous relationship between $q_A(t)$ and $v_{AN}(t)$ is as given in Eq. (2.5).

$$v_{AN}(t) = V_d \, q_A(t) \qquad (2.5)$$

Considering that V_d is a constant DC voltage, the CCA value of $v_{AN}(t)$ can be obtained as given in Eq. (2.6).

$$\bar{v}_{AN}(t) = \bar{q}_A(t) \, V_d = d(t) \, V_d \qquad (2.6)$$

Equivalently, $\bar{v}_{AN}(t)$ can also be obtained by applying the definition of CCA to its instantaneous waveform as shown in Fig. 2.5, resulting in the same expression as given in Eq. (2.6).

2.2.2 Pulse-Width Modulation of a Power Pole

A popular pulse-width modulation (PWM) process to generate the switching signal $q_A(t)$ is the carrier-based approach. In this method a control voltage, $v_{cA}(t)$ referred to as the modulating signal is compared with a triangular carrier voltage, $v_{tri}(t)$, and the points of intersection of these two voltages determine the switching instants of the power pole. The PWM process is defined in Eq. (2.7) and is illustrated in Fig. 2.6.

$$\begin{aligned} q_A(t) &= 1 \text{ for } v_{cA}(t) \geq v_{tri}(t) \\ q_A(t) &= 0 \text{ for } v_{cA}(t) < v_{tri}(t) \end{aligned} \qquad (2.7)$$

Figure 2.6a shows the triangular wave with a peak value of $\hat{V}_{tri}$ compared with a control voltage, which in this example is a sinusoidal voltage. The frequency of the triangular carrier should be much higher than that of the sinusoidal modulating

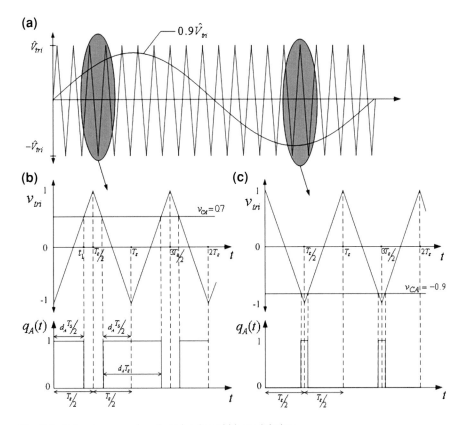

Fig. 2.6 Triangle comparison-based pulse-width modulation

signal. The ratio of the carrier frequency to the modulating frequency is called the frequency modulation index and is usually a high value, for example, above 100. Figures 2.6b and c show an expanded view of a small duration in Fig. 2.6a, corresponding to two switching periods each in the positive and negative portions, respectively, of the sinusoidal waveform, in order to clearly show the comparison and the resulting PWM waveforms. This is also used to derive the relationship between the control voltage, the duty ratio, and the pole output voltage.

Referring to Fig. 2.6b, the rising portion of the triangular waveform is expressed as

$$v_{\text{tri}}(t) = -\hat{V}_{\text{tri}} + \frac{2\hat{V}_{\text{tri}}}{\left(\dfrac{T_s}{2}\right)} t \quad 0 \le t \le \frac{T_s}{2} \qquad (2.8)$$

The time t_1 at which $v_{\text{tri}}(t)$ becomes equal to v_{cA} can be obtained from Eq. (2.8) by substituting $t = t_1$ and $v_{\text{tri}} = v_{cA}$

$$-\hat{V}_{\text{tri}} + \frac{2\hat{V}_{\text{tri}}}{\left(\dfrac{T_s}{2}\right)} t_1 = v_{cA} \tag{2.9}$$

$$t_1 = \left(\hat{V}_{\text{tri}} + v_{cA}\right) \frac{\left(\dfrac{T_s}{2}\right)}{2\hat{V}_{\text{tri}}} \tag{2.10}$$

From Fig. 2.6b, t_1 is the duration for which $q_A = 1$ in one half of the switching period, $T_s/2$. Using the definition given in Eq. (2.3), the duty ratio $d(t)$ can be derived using Eq. (2.10) and as given in Eq. (2.11), which is an important relationship between the duty ratio and the control voltage of a power pole.

$$d(t) = \frac{t_1}{\left(\dfrac{T_s}{2}\right)} = \frac{\left(\hat{V}_{\text{tri}} + v_{cA}(t)\right)}{2\hat{V}_{\text{tri}}} = \frac{1}{2} + \frac{1}{2\hat{V}_{\text{tri}}} v_{cA}(t) \tag{2.11}$$

Using Eqs (2.6) and (2.11), the important relationship between the control voltage and the cycle-by-cycle average value of the output voltage of the power pole can be obtained as given in Eq. (2.12).

$$\bar{v}_{AN}(t) = d(t) V_d = \frac{V_d}{2} + \frac{V_d}{2\hat{V}_{\text{tri}}} v_{cA}(t) \tag{2.12}$$

For the sake of simplicity and without loss of generality, the peak of the triangular voltage, $\hat{V}_{\text{tri}}$, is assumed to be unity in the analysis in the following sections of this book. Under this assumption, the control voltage is constrained to be $-1 \leq v_{cA}(t) \leq 1$, and the relationship between control voltage and average pole output voltage becomes

$$\bar{v}_{AN}(t) = \frac{V_d}{2} + \frac{V_d}{2} v_{cA}(t) \tag{2.13}$$

As seen from Eq. (2.13), the power pole *amplifies* the control voltage v_{cA} by a factor of $\frac{V_d}{2}$ and also adds a DC offset of $\frac{V_d}{2}$. This relationship is valid under all static and dynamic conditions provided the frequency of the triangular carrier is well above (at least an order of magnitude) the highest frequency component of the control voltage.

Example 2.1 illustrates the concept of the power pole as an amplifier using pulse-width modulation, and the different components of the output voltage of a power pole.

Example 2.1 Consider the power pole shown in Fig. 2.7 with a DC link voltage $V_d = 100$ V and the control voltage given by $v_{cA}(t) = 0.7 \sin(2\pi \times 50t) + 0.2 \sin(2\pi \times 150t)$ V. The switching frequency (or the frequency of the carrier triangular waveform) is given as 3 kHz. The peak of the triangular wave $\hat{V}_{\text{tri}}$ can be assumed to be 1 V.

(a) Calculate the CCA value of the pole output voltage, $\bar{v}_{AN}(t)$
(b) Plot $v_{cA}(t)$, $v_{AN}(t)$ and $\bar{v}_{AN}(t)$
(c) Plot the frequency spectrum of $v_{AN}(t)$

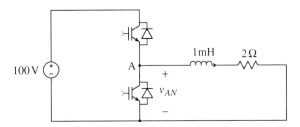

Fig. 2.7 Power pole corresponding to Example 2.1

(a) The CCA value of the pole output voltage is given by

$$\bar{v}_{AN}(t) = \frac{V_d}{2} + \frac{V_d}{2} \, v_{cA}(t) = 50 + 35 \sin(2\pi \times 50t) + 10 \, \sin(2\pi \times 150t) \, \text{V} \quad (2.14)$$

(b) The plots of $v_{cA}(t)$, $v_{AN}(t)$ and $\bar{v}_{AN}(t)$ are obtained using a simulation tool
    called PLECS (Piece-wise Linear Electric Circuit Simulation) and are shown
    in Fig. 2.8. Note that the instantaneous $v_{AN}(t)$ switches between V_d and 0 at
    the switching frequency of 3 kHz, and the CCA value $\bar{v}_{AN}(t)$ has 50 V DC
    offset and amplifies $v_{cA}(t)$ by a gain of 50 as derived in Eq. (2.14). The
    waveform of $\bar{v}_{AN}(t)$ will be much smoother if the switching frequency is
    higher. The choice of 3 kHz was made in the example in order to make the
    individual pulses of the PWM waveform clearer to view.

(c) The frequency spectrum of $v_{AN}(t)$ is also obtained using PLECS and is shown
    in Fig. 2.9. As seen, the frequency spectrum has a 50 V DC component, 50 Hz
    component with 35 V peak, and a 150 Hz component with 10 V peak. The
    next dominant frequency component is the switching frequency (3 kHz) along
    with its side bands.

2.2.3 Pole Current and Analysis of DC Link Current

The current through the pole is required to be a non-switching current. The pole
current is usually a current through an inductor or through the winding of a motor,
with finite (and usually low) rate of change determined by the pole voltage.
A typical waveform of the pole current, $i_A(t)$ for many power converters is as
shown in Fig. 2.10. In DC–DC applications, the pole current is predominantly DC,
with small switching frequency (and its harmonics) component; in DC–AC or

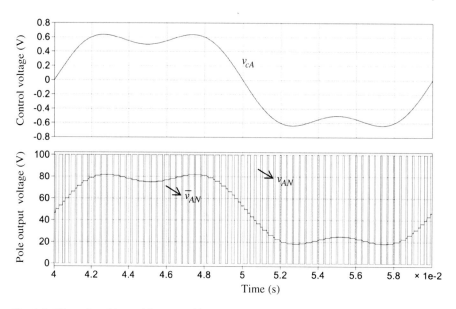

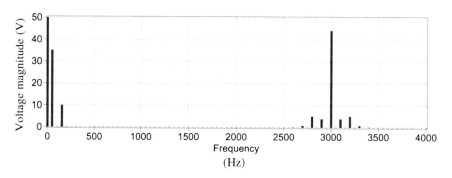

Fig. 2.8 Plots of $v_{cA}(t)$, $v_{AN}(t)$ and $\bar{v}_{AN}(t)$ corresponding to Example 2.1

Fig. 2.9 Frequency spectrum of pole output voltage corresponding to Example 2.1

AC–DC applications, it is predominantly low-frequency AC, with small switching frequency (and its harmonics).

In this section, we analyze the DC current drawn by the DC terminals of the power pole, denoted $i_{dA}(t)$ in Fig. 2.4 in terms of the pole current, and we derive an expression for its CCA value. This will lead to the average model of the power pole described in Sect. 2.2.4. From Fig. 2.4 we can see that $i_{dA} = i_A$ when $q_A = 1$, and $i_{dA} = 0$ when $q_A = 0$. This is also illustrated in Fig. 2.10. Therefore, the instantaneous $i_{dA}(t)$ can be expressed as in Eq. (2.15).

$$i_{dA}(t) = q_A(t)\, i_A(t) \tag{2.15}$$

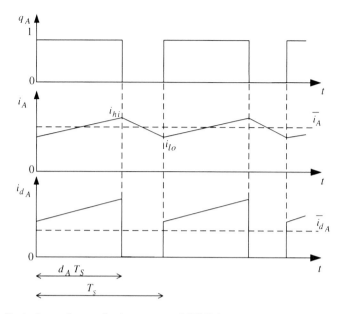

Fig. 2.10 Typical waveforms of pole current and DC link current

In order to derive the CCA value of i_{dA}, we make a valid assumption that within a switching period, the average value of i_A over the interval when $q_A = 1$ is equal to the average value of i_A over the interval where $q_A = 0$. For the typical piecewise linear pole current waveforms as shown in Fig. 2.10 this is satisfied if the condition given in Eq. (2.16) is met.

$$i_A(t + T_s) = i_A(t) \tag{2.16}$$

The above requirement is strictly met in DC steady state for DC–DC converters. For other types of conversion such as DC to sinusoidal AC applications, and during transients in all applications, this requirement may not be strictly met. However, with switching frequencies well above the highest frequency component of the modulating waveform, and well above the control bandwidth of the various control loops, Eq. (2.16) is a good approximation ($i_A(t + T_s) \approx i_A(t)$), and will be made use of in further analysis.

From Fig. 2.10 and with the assumption of Eq. (2.16), we can see that

$$\bar{i}_{dA}(t) = d_A(t)\left(\frac{i_{lo} + i_{hi}}{2}\right) = d_A(t)\bar{i}_A(t) \tag{2.17}$$

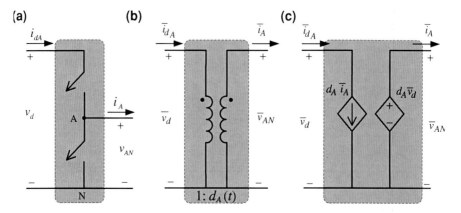

Fig. 2.11 Average modeling of a power pole: **a** switch model, **b** average model using an ideal transformer, **c** average model with the ideal transformer implemented with controlled voltage and current sources

2.2.4 Average Model of a Power Pole

The complete average model of a power pole is then given by Eqs. (2.6) and (2.17). These two equations are similar to those of an ideal transformer with its turns ratio equal to the instantaneous duty ratio. Figure 2.11 illustrates this concept of modeling the power pole in an average sense as an ideal transformer [1, 2]. It may be noted that this is only a mathematical concept useful for analysis and control design, and a power pole cannot be replaced by a physical transformer. In particular, the power pole differs from a physical transformer in that the turns ratio of the power pole model is continuously variable and DC voltages can be applied to the terminals.

Example 2.2 Draw the average model of the power pole considered in Example 2.1 including the control voltage $v_{cA}(t) = 0.7 \sin(2\pi \times 50t) + 0.2 \sin(2\pi \times 150t)$ V. Simulate the circuit using the switch model and the average model and compare the waveforms of $i_A(t)$ obtained from the switch model, $\bar{i}_A(t)$ which is obtained by taking the using cycle-by-cycle average value of $i_A(t)$ and $\bar{i}_A(t)$ obtained directly from the average model simulation.

The average model of the power pole and the system of Example 2.1 is shown in Fig. 2.12. The generation of the duty ratio $d(t)$ from the control voltage $v_{cA}(t)$ is also shown explicitly in the figure.

$$d(t) = \frac{1}{2} + \frac{1}{2}v_{cA}(t) = 0.5 + 0.35 \sin(2\pi \times 50t) + 0.1 \sin(2\pi \times 150t) \text{ V}$$

The system was simulated in PLECS using ideal switches. The calculation of the CCA values from the instantaneous values is derived in Eq. (2.18), and the block diagram of the CCA implementation in PLECS is shown in Fig. 2.13a.

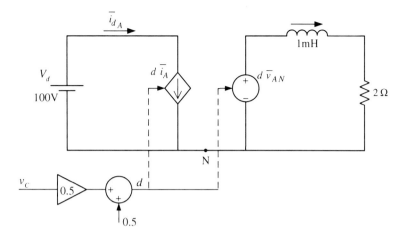

Fig. 2.12 Average model of the power converter of Example 2.1

$$\bar{x}(t) = \frac{1}{T_s} \int\limits_{t-T_s}^{t} x(\tau)\, d\tau = \frac{1}{T_s} \left[\int\limits_{0}^{t} x(\tau)\, d\tau - \int\limits_{0}^{t-T_s} x(\tau)\, d\tau \right] \tag{2.18}$$

In Eq. (2.18), $\int\limits_{0}^{t-T_s} x(\tau)\, d\tau$ can be obtained either as $\int\limits_{0}^{t} x(\tau - T_s)\, d\tau$ or as $I_x(t - T_s)$ where, $I_x(t) = \int\limits_{0}^{t} x(\tau)\, d\tau$.

The simulation plots of $i_A(t)$ and $\bar{i}_A(t)$ obtained from the average model and from switching model (using Eq. (2.18)) are shown superimposed in Fig. 2.13b. The plots corresponding to the switching model are obtained at switching frequencies of 10 kHz. As seen, the results of the average model match very closely with the switching model, confirming the validity of the average model even when the pole current is not a constant DC current.

2.3 Single Pole Converter

A single power pole by itself is a practical power converter capable of a two-quadrant operation (bidirectional current but unipolar voltage at the current stiff port). Single pole converters find widespread applications in many DC-DC converters and DC motor drives. These converters have bidirectional power flow capability, i.e., the power flow can be from the voltage stiff port to the current stiff port or vice versa depending on the operating conditions and control voltages. The voltage stiff port is always at a higher voltage (and lower current) than the current

(a)

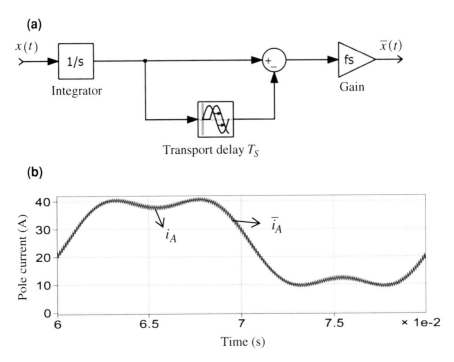

(b)

Fig. 2.13 **a** Implementation of cycle-by-cycle averaging in simulation and **b** pole current for Example 2.2 obtained using switch model, cycle-by-cycle averaging in switch model, and from average model

Fig. 2.14 A single pole converter used in a DC motor drive application

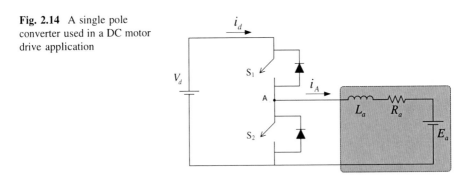

stiff port regardless of the power flow direction. When the voltage stiff port has a source connected to it, and the power flow is toward the current port, it corresponds to the buck mode of operation. When the current stiff port has the source, and the power flow is toward the voltage stiff port, the operation is termed as the boost mode.

A single pole converter used in DC-DC power conversion applications, specifically for a two-quadrant DC motor drive is shown in Fig. 2.14. The components

Fig. 2.15 Average model of
a single pole converter used
in DC motor drive

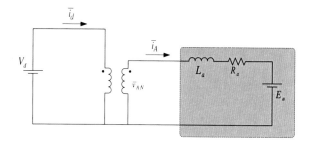

Fig. 2.16 Equivalent circuit
for calculating high-
frequency ripple

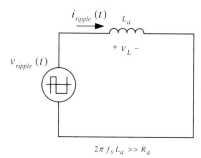

inside the shaded box— L_a, R_a, and E_a correspond to the model of the armature
circuit of a DC motor, where L_a and R_a are the armature inductance and resistance,
respectively, and E_a is the induced electromotive force (emf) in the armature
winding.

The complete analysis of a power converter leading to the design of various
power stage components involves two distinct parts—average (CCA) analysis and
switching frequency analysis. For average analysis, we can make use of the
average model (ideal transformer) of the power pole as shown in Fig. 2.15. Steady-
state as well as low-frequency (compared to switching frequency) transients can be
studied using the average model (see Example 2.3 on page 37).

The switching frequency analysis is required for understanding the requirements on the filter components, to determine power quality metrics, for selection
of voltage and peak current ratings for various devices, and to analyze power
losses. For example, the high frequency equivalent circuit of Fig. 2.16 can be used
to determine the switching frequency ripple in the pole current (or armature current). The voltage source v_{ripple} corresponds to the high frequency component of
the pole voltage v_{AN}, and i_{ripple} corresponds to the high frequency component of
the pole current.

$$v_{\text{ripple}}(t) = v_{AN}(t) - \bar{v}_{AN}(t)$$
$$i_{\text{ripple}}(t) = i_A(t) - \bar{i}_A(t)$$

$$(2.19)$$

In the equivalent circuit of Fig. 2.16 the resistance is assumed to be negligible
in comparison with the impedance of the inductance at the switching frequency

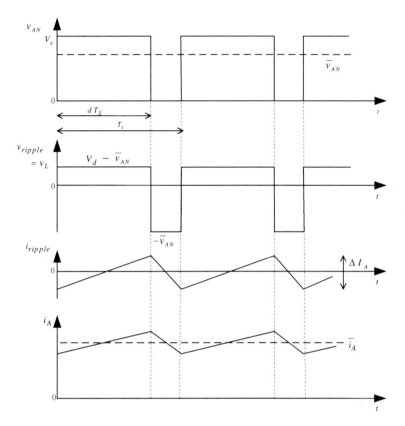

Fig. 2.17 High-frequency ripple voltage and current in a single pole converter

and its harmonics, and the voltage source E_a is shorted since it is a DC source and has no high frequency component. Figure 2.17 shows the relevant waveforms of the equivalent circuit and can be used to calculate the peak–peak ripple (Δi_A) in the pole current.

During the ON interval, i.e., with $q_A = 1$,

$$L_a \frac{di_A}{dt} = v_L(t) = V_d - \bar{v}_{AN} \tag{2.20}$$

Therefore, the peak–peak ripple is obtained as

$$\Delta i_A = \frac{(V_d - \bar{v}_{AN}) \, d \, T_s}{L_a} \tag{2.21}$$

For a given requirement on the peak–peak ripple in the pole current, the requirements on the filter inductance and the switching frequency can be obtained from Eq. (2.21). Also, note that from Fig. 2.17 the complete waveform for the pole current can be obtained by the superposition of the average current obtained from the average model of Fig. 2.15, and the ripple current obtained from the high frequency model of Fig. 2.16.

Example 2.3 Consider the single pole converter shown in Fig. 2.14 with the parameter values given as $L_a = 500\,\mu\mathrm{H}$, $R_a = 1\Omega$, $E_a = 80\,\mathrm{V}$ and $V_d = 100\,\mathrm{V}$. Calculate the required duty ratio d_A, and the control voltage v_{cA} in steady state for two different cases to obtain—(1) $\bar{i}_A = 10\,\mathrm{A}$ and (2) $\bar{i}_A = -10\,\mathrm{A}$. Calculate the average input DC current for each case. Also, calculate the peak–peak ripple in the inductor current for each of the two cases above. The switching frequency is 20 kHz.

Since we are considering DC steady state, the inductor can be assumed short circuited (note that in DC steady state with switch mode converters the instantaneous voltage across the inductor is not zero, but a large positive or negative voltage; only the average (CCA) value of the voltage is zero, justifying the short circuit in the average model).

Case (1): Using the average model shown in Fig. 2.15,

$$\bar{v}_{AN} = E_a + \bar{i}_A R_a = 80 + 10 \times 1 = 90\,\mathrm{V}$$

$$d_A = \frac{v_{AN}}{V_d} = \frac{90}{100} = 0.9$$

From Eq. (2.11) and with $\hat{V}_{tri} = 1\,\mathrm{V}$, $v_{cA} = (d_A - 0.5)2 = 0.8\,\mathrm{V}$
From the average model, $\bar{i}_{dA} = d_A \bar{i}_A = 0.9 \times 10 = 9\,\mathrm{A}$
The peak–peak ripple is calculated using Eq. (2.21) as

$$\Delta i_A = \frac{(V_d - \bar{v}_{AN})\,d\,T_s}{L_a} = \frac{(100 - 90) \times 0.9 \times 50 \times 10^{-6}}{500 \times 10^{-6}} = 0.9\,\mathrm{A}$$

Case (2):

$$\bar{v}_{AN} = E_a + \bar{i}_A R_a = 80 + (-10) \times 1 = 70\,\mathrm{V}$$

$$d_A = \frac{v_{AN}}{V_d} = \frac{70}{100} = 0.7$$

$$v_{cA} = (d_A - 0.5)2 = 0.4\,\mathrm{V}$$

$$\bar{i}_{dA} = d_A \bar{i}_A = 0.7 \times (-10) = -7\,\mathrm{A}$$

$$\Delta i_A = \frac{(V_d - \bar{v}_{AN})\,d\,T_s}{L_a} = \frac{(100 - 70) \times 0.7 \times 50 \times 10^{-6}}{500 \times 10^{-6}} = 2.1\,\mathrm{A}$$

Example 2.4 The single pole converter of Example 2.3 is driven by a duty ratio shown in Fig. 2.18 (note that the two values correspond to those calculated in Example 2.3 for $\bar{i}_A$ of 10 A and -10 A). Derive an expression for the current $\bar{i}_A(t)$

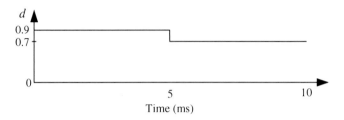

Fig. 2.18 Duty ratio for Example 2.4

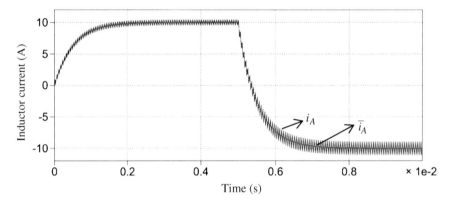

Fig. 2.19 Inductor current of Example 2.4 from both switch model simulation and analytical derivation

assuming the initial inductor current to be zero. Verify the derived expression with the simulated inductor current from the complete switching model.

Using Laplace transforms in the average model of Fig. 2.15,

$$I_A(s) = \frac{V_{AN}(s) - E_a(s)}{R_a + s\,La}$$

$$v_{AN}(t) = V_d\,d(t) = 90\,u(t) - 20\,u(t - 0.005)\ \text{V}$$

$$V_{AN}(s) = \frac{90}{s} - \frac{20}{s}e^{-s0.005}\ \text{V};\ E_a(s) = \frac{80}{s}\ \text{V}$$

Substituting the expressions for $V_{AN}(s)$ and $E_a(s)$ and using partial fractions,

$$I_A(s) = \frac{10}{s\,(R_a + s\,La)} - \frac{20\,e^{-s0.005}}{s\,(R_a + s\,La)}\ \text{A}$$

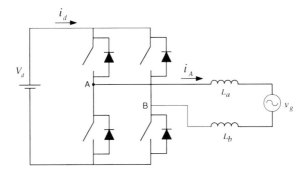

Fig. 2.20 A two-pole converter used in a DC–AC (or AC–DC) application

Taking inverse Laplace transform,

$$\bar{i}_A(t) = 10\left(1 - e^{-t/\tau}\right)u(t) - 20\left(1 - e^{-(t-0.005)/\tau}\right)u(t - 0.005)\,\text{A}$$

where $\tau = \frac{L_a}{R_a} = 500 \times 10^{-6}\,s$

The single pole converter with given parameters and duty ratio was simulated using the switching model in PLECS. The instantaneous inductor current obtained from this simulation is superimposed with the average current expression derived above and is shown in Fig. 2.19. As seen, the waveform from the analytical derivation matches very well with the average of the inductor current from the switch model even during transients. The plots also verify the value of peak–peak ripple current derived in Example 2.3.

2.4 Two-Pole Converter

Two-pole converters, also known as full-bridge converters, are widely used in single-phase DC–AC or AC–DC power conversion applications and in DC motor drives. A major application area is the integration of solar photovoltaic (PV) energy to the single-phase AC distribution system. Two-pole converters are capable of four-quadrant operation, and power flow between the AC and DC sides can be bidirectional.

The schematic of a two-pole, full-bridge converter is shown in Fig. 2.20 corresponding to a grid-connected DC–AC operation (inverter) or AC–DC operation (rectifier). The inverter mode corresponds to power flow from the DC side to the AC side, and the rectifier mode corresponds to power flow from the AC side to the DC side (for example, to charge a battery at the DC terminals). The power at the AC side can be injected or absorbed at any desired power factor. The same converter topology can be used for stand-alone DC–AC applications (i.e., without a grid connection) with additional capacitive filters at the AC terminals.

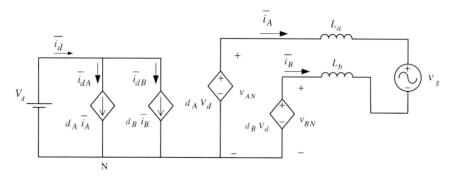

Fig. 2.21 Average model of a two-pole converter

2.4.1 Average Model of a Two-Pole Converter

The analysis of a power pole discussed in detail in the earlier sections can be readily used for the complete analysis of the two-pole converter. Once the control voltages for each of the power poles are known, the analysis of the individual pole is identical to that of the single power pole using the average model. The average model of the two-pole converter in Fig. 2.20 can be obtained by using two ideal transformers corresponding to the two poles as shown in Fig. 2.21. A simplified average model is also derived later in Sect. 2.4.2.

2.4.2 Unipolar PWM

Different types of PWM schemes are possible for a two-pole converter based on how power pole B is driven with respect to power pole A. Some of the schemes used in practical converters include unipolar PWM (discussed in detail below), bipolar PWM where the switching signals for the two poles are related as $q_B(t) = 1 - q_A(t)$, and a third scheme where one of the poles is driven at the switching frequency with variable duty ratio, and the other pole driven at the fundamental frequency with a fixed duty ratio of 0.5. The average model shown in Fig. 2.21 is valid for all the schemes.

The unipolar PWM has several advantages including a desirable frequency spectrum for the PWM pole voltages, output current, and DC current. The dominant high frequency component in unipolar PWM is at twice the switching frequency, which is a significant advantage in terms of lower filter requirement compared to other PWM schemes where the dominant high frequency component is at the switching frequency. Hence, the unipolar PWM is widely used at present, and further discussions in this chapter will focus only on this PWM method.

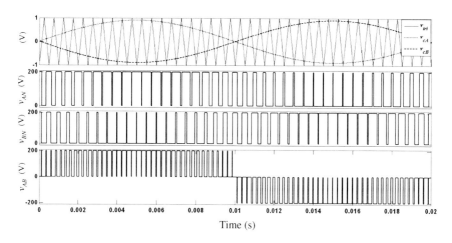

Fig. 2.22 Waveforms of two-pole converter with unipolar PWM

In unipolar PWM, two modulating (or control) signals are compared with a common triangular carrier wave (we assume $\hat{V}_{\text{tri}} = 1\ V$). The definition of unipolar PWM in terms of the two control voltages for the two poles is given in Eq. (2.22).

$$v_{cB}(t) = -v_{cA}(t) \tag{2.22}$$

In practice, a control voltage $v_c(t)$ is generated by a controller based on the control objectives and measured values. This is directly used as the modulating signal for one of the poles, say A, and its negative value is used as the modulating signal for pole B, as indicated in Eq. (2.23).

$$v_c(t) = v_{cA}(t) = -v_{cB}(t) \tag{2.23}$$

Figure 2.22 shows the salient waveforms of a two-pole converter employing a unipolar PWM scheme. As seen, the instantaneous terminal voltage, v_{AB}, is always positive (or zero) in the positive half-cycle of the modulating signal, and always negative (or zero) in the negative half-cycle. The term unipolar PWM refers to this characteristic of the PWM waveform.

Using the relationship between duty ratio and control voltage given in Eq. (2.11) with the assumption of $\hat{V}_{\text{tri}} = 1\ V$, and Eq. (2.23), we obtain

$$d_A(t) = \frac{1}{2} + \frac{1}{2} v_c(t)$$
$$d_B(t) = \frac{1}{2} - \frac{1}{2} v_c(t) \tag{2.24}$$

We define an equivalent duty ratio $d(t)$ for the combined two-pole converter as given in Eq. (2.25) in order to further simplify the model and analysis of a two-pole converter.

$$d(t) = d_A(t) - d_B(t) = \frac{1}{2} + \frac{1}{2}v_c(t) - \frac{1}{2} + \frac{1}{2}v_c(t) = v_c(t) \qquad (2.25)$$

The corresponding expressions for the pole voltages are given in Eqs. (2.26) and (2.27).

$$\bar{v}_{AN}(t) = d_A(t) V_d = \frac{V_d}{2} + \frac{V_d}{2}v_c(t) \qquad (2.26)$$

$$\bar{v}_{BN}(t) = d_B(t) V_d = \frac{V_d}{2} - \frac{V_d}{2}v_c(t) \qquad (2.27)$$

From Eqs. (2.26) and (2.27), the two-pole terminal voltage (v_{AB}) is obtained as given in Eq. (2.28).

$$\bar{v}_{AB}(t) = \bar{v}_{AN}(t) - \bar{v}_{BN}(t) = V_d v_c(t) = V_d d(t) \qquad (2.28)$$

Hence, the two-pole converter amplifies the control voltage v_c by a gain equal to the voltage of the voltage stiff port, V_d. Fig. 2.23 shows the expanded waveforms of a two-pole converter with unipolar PWM over two switching periods. The average values of different pole and terminal voltages are highlighted and are consistent with the expressions derived above. The doubling of frequency in the terminal voltage ($2f_s$) compared to the pole voltage (f_s) is also clear from the figure.

The DC current drawn by the voltage stiff port is the sum of the DC currents drawn by the individual poles.

$$\bar{i}_d(t) = \bar{i}_{dA}(t) + \bar{i}_{dB}(t) = d_A(t)\bar{i}_A(t) + d_B(t)\bar{i}_B(t) \qquad (2.29)$$

Noting that $i_A(t) = -i_B(t)$, and $d_A(t) - d_B(t) = d(t) = v_c(t)$

$$\bar{i}_d(t) = v_c(t)\,\bar{i}_A(t) = d(t)\,\bar{i}_A(t) \qquad (2.30)$$

The combined average model of the two-pole converter in terms of the terminal quantities is given by Eqs. (2.28) and (2.30). This is the model of an ideal transformer with turns ratio equal to the instantaneous $d(t)$ [1, 2]. The combined average model of the two-pole converter is shown in Fig. 2.24.

2.4.3 High Frequency Ripple with Unipolar PWM

Similar to the single pole converter case, the complete analysis of a two-pole converter can also be done by superposition of fundamental frequency equivalent circuit and high frequency equivalent circuit. The fundamental frequency analysis is illustrated later through Example 2.5. In this section, we are interested in the high frequency analysis. Referring to the schematic of Fig. 2.20, the entire high-frequency ripple, $v_{\text{ripple}}(t)$ in the terminal voltage $v_{AB}(t)$ as defined in Eq. (2.31) is

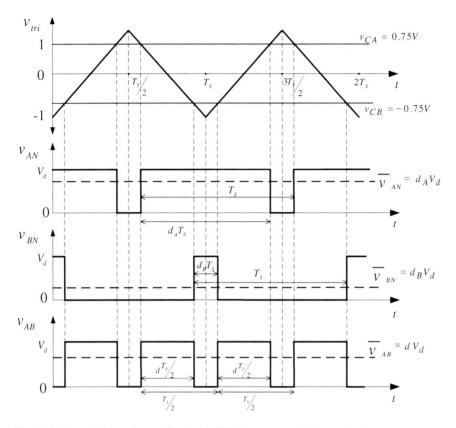

Fig. 2.23 Expanded waveforms of unipolar PWM over two switching periods

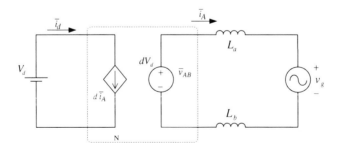

Fig. 2.24 Combined (single transformer) average model of the two-pole converter

applied across the series inductor $L = L_a + L_b$ leading to the high frequency ripple in the inductor current.

$$v_{ripple}(t) = v_{AB}(t) - \bar{v}_{AB}(t) = v_{AB}(t) - V_d\, d(t) \qquad (2.31)$$

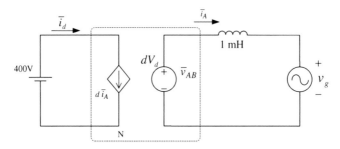

Fig. 2.25 Average model corresponding to Example 2.5

The peak–peak ripple in the inductor current in the kth switching cycle is given by Eq. (2.32) where $d(k)$ is the combined duty ratio in the kth switching cycle.

$$\Delta i_A(k) = \frac{(V_d - V_d d(k))d(k)\frac{T_s}{2}}{L} \tag{2.32}$$

Since the duty ratio varies over a wide range over the duration of the fundamental period, the peak–peak ripple in the inductor current also varies correspondingly. For detailed analysis of high-frequency ripple and for design of the filters to meet a given limit on waveform distortion in two-pole converters, the interested reader is referred to [3].

Example 2.5 Consider the two-pole converter shown in Fig. 2.20 with parameter values $L_a = L_b = 500\,\mu\text{H}$, $V_d = 400\,\text{V}$, and the grid voltage $v_g = 340\,\sin(377t)$ V. It uses unipolar PWM with a switching frequency of 20 kHz. Calculate the required control voltage $v_c(t)$ for the following two cases. Also, calculate the DC currents in each case. Verify through simulation that the control voltages calculated result in the right grid current for the required power flow, and the calculated DC current.

Case 1 The power flow is required to be 5 kW *from the DC source to the grid* at 0.866 lagging power factor (i.e., i_g as defined in Fig. 2.20 lags v_g by 30°).

Case 2 The power flow is required to be 5 kW *from the grid to the DC source* at unity power factor.

The average model for the given converter and parameter values is shown in Fig. 2.25 and is valid for both operating conditions corresponding to inverter and rectifier modes of operation, respectively. Since we are interested only in the fundamental quantities in steady state while calculating the control voltage v_c, it is convenient to use phasor analysis (bold upper case letters denote phasors).

Case 1 (inverter mode):
The peak value of the required current can be calculated from

$$P_g(t) = \frac{\hat{V}_g \hat{I}_g}{2}\cos\theta \Rightarrow \hat{I}_g = \frac{2 \times 5000}{340 \times 0.866} = 33.96\,\text{A}$$

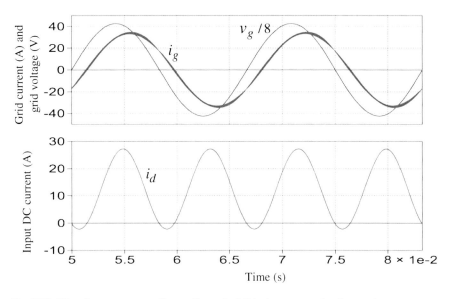

Fig. 2.26 Waveforms corresponding to Example 2.5 in inverter mode of operation

$$V_g = 340 \angle 0° \text{ V and } I_g = 33.96 \angle -30° \text{ A}$$

$$V_{AB} = V_g + V_L = V_g + j\omega L I_L = 340 \angle 0° + 12.8 \angle 60°$$
$$= 346.58 \angle 1.834° \text{ V}$$

$$V_c = \frac{V_{AB}}{V_d} = 0.86645 \angle 1.834° \text{ V}$$
$$\therefore v_c(t) = 0.86645 \ \sin(377t + 1.834°) \text{ V}$$

From Fig. 2.25, the input DC current is given by $i_d(t) = v_c(t) i_g(t)$

$$i_d(t) = 29.425 \ \sin(377t + 1.834°) \ \sin(377t - 30°) \text{ A}$$
$$= 12.499 - 12.97 \cos(754t) - 6.944 \sin(754t) \text{ A}$$

The switch model of the two-pole converter was simulated in PLECS with the calculated value of $v_c(t)$.. The resulting grid current waveform and the DC current waveform from simulation are shown in Fig. 2.26 and are superimposed with the expected values, which match well.

Case 2 (rectifier mode):

The peak value of the required current can be calculated from

$$P_g(t) = \frac{\hat{V}_g \hat{I}_g}{2} \cos\theta \Rightarrow \hat{I}_g = \frac{2 \times 5000}{340} = 29.41 \text{ A}$$

$$V_g = 340 \angle 0° \text{ V and } I_g = 29.41 \angle 180° \text{ A}$$

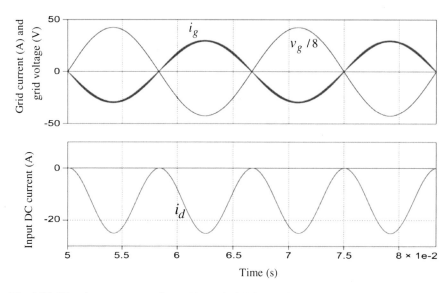

Fig. 2.27 Waveforms corresponding to Example 2.5 in rectifier mode of operation

$$V_{AB} = V_g + V_L = V_g + j\omega L I_L = 340 - j11.088$$
$$= 340.181 \angle -1.868° \text{ V}$$

$$V_c = \frac{V_{AB}}{V_d} = 0.8505 \angle -1.8679° \text{ V}$$
$$\therefore v_c(t) = 0.8505 \sin(377t - 1.8679°) \text{ V}$$

From Fig. 2.25, the input DC current is given by $i_d(t) = v_c(t) i_g(t)$

$$i_d(t) = 25.01 \sin(377t - 1.8679°) (-\sin(377t)) \text{ A}$$
$$= -12.5 + 12.5 \cos(754t - 1.868°) \text{A}$$

The switch model of the two-pole converter was simulated in PLECS with the calculated value of $v_c(t)$. The resulting grid current waveform and the DC current waveform from simulation are shown in Fig. 2.27 and are superimposed with the expected values, which match well.

2.5 Three-Pole Converters for Three-Phase Applications

The three-pole converter shown in Fig. 2.28 is a popular topology in numerous DC–AC and AC–DC applications, and can even be called the 'work-horse' in renewable energy integration and motor drive applications. It is presently the most

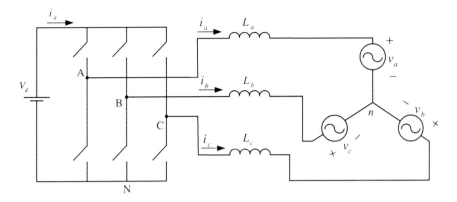

Fig. 2.28 A three-pole converter used in three-phase applications

widely used converter topology in the grid interface of wind energy, applied in both Type 3 and Type 4 wind generator systems.

Most applications of a three-pole converter involve the synthesis of low-frequency (in relation to the switching frequency) three-phase AC voltages from the DC link. The power flow direction can be from the AC source to the DC link or vice versa depending on the application and can change dynamically. In motor drive applications, the fundamental frequency needs vary over a wide range—from almost zero to more than 100 Hz; in PV to grid applications, the fundamental frequency is the grid frequency of 60 Hz or 50 Hz; and in wind integration, the fundamental frequency of the rotor side converter can vary from a few Hz to a few tens of Hz.

In conventional sine-triangle comparison-based PWM, a controller generates the three modulating signals, one for each phase, at the required fundamental frequency, and with a phase-shift of 120° during balanced, steady-state conditions, as given in Eq. (2.33). These three signals are compared with a common triangular waveform at the switching frequency to obtain the PWM signal for each pole. The process of PWM generation for the three-pole converter, the resulting pole voltages, and line–line voltages are illustrated in Fig. 2.29.

$$v_{cA}(t) = V_c \cos(\omega t) \text{ V}$$
$$v_{cB}(t) = V_c \cos(\omega t - 120°) \text{ V} \qquad (2.33)$$
$$v_{cC}(t) = V_c \cos(\omega t - 240°) \text{ V}$$

where ω is the fundamental frequency.

The average model analysis developed for a single pole can be readily applied here. Similar to Eq. (2.13) for a single pole, Eq. (2.34) can be derived for individual pole voltages for the three-pole converter. We assume that $\hat{V}_{tri} = 1 \text{ V}$ and $V_c \leq 1 \text{ V}$.

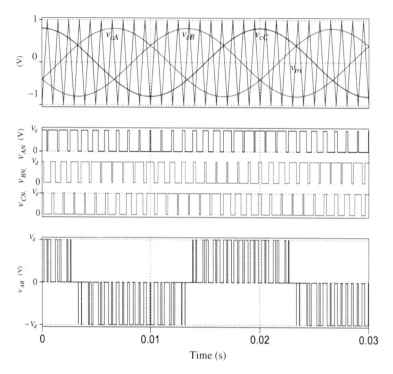

Fig. 2.29 Three-phase PWM using sine-triangle comparison

$$\bar{v}_{AN}(t) = \frac{V_d}{2} + \frac{V_d}{2} V_c \cos(\omega t) \text{ V}$$

$$\bar{v}_{BN}(t) = \frac{V_d}{2} + \frac{V_d}{2} V_c \cos(\omega t - 120°) \text{ V} \qquad (2.34)$$

$$\bar{v}_{CN}(t) = \frac{V_d}{2} + \frac{V_d}{2} V_c \cos(\omega t - 240°) \text{ V}$$

The average line–line voltages can then be directly obtained from Eq. (2.34) by taking the difference of the respective pole voltages. For example, the average line–line voltage $\bar{v}_{AB}$ is given in Eq. (2.35). As seen, the DC component in the individual pole voltages is cancelled in the line–line voltage, and the highest peak line–line voltage (average) that can be generated from a three-pole converter is $\frac{\sqrt{3}}{2} V_d$.

$$\bar{v}_{AB}(t) = \frac{V_d}{2} V_c \left[\cos(\omega t) - \cos(\omega t - 120°)\right]$$

$$= \frac{\sqrt{3}}{2} V_d V_c \cos(\omega t + 30°) \text{ V} \qquad (2.35)$$

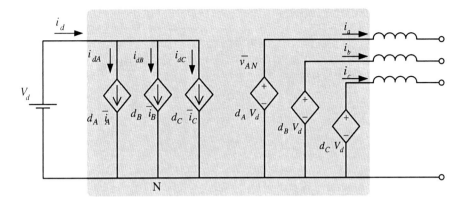

Fig. 2.30 Average model of a three-pole converter

The complete average model of the three-pole converter can be derived from the average model of a single pole and is shown in Fig. 2.30. The current drawn from the DC link is the sum of the individual currents drawn by the three poles. It has a DC component and a switching frequency component and its harmonics, but no low frequency component or its harmonics (such as twice the line frequency component in single-phase converters). The CCA waveform of the DC link current, therefore, will only have a DC value and is equal to the power processed divided by the DC link voltage magnitude. Hence, the size of capacitive filter needed at the DC link is significantly lower than single-phase converters.

It may be noted that an alternate method of PWM generation called the space vector PWM (SVPWM) is another very popular technique used in three-phase applications. SVPWM offers the advantages of higher DC bus utilization (about 15% higher), and reduced high frequency ripple under certain operating conditions [4]. However, detailed discussion on SVPWM is beyond the scope of this book. It may also be noted that PWM waveforms similar to those of SVPWM can also be generated by sine-triangle comparison methods discussed here by suitable addition of triplen harmonics in the modulating waveforms. The triplen harmonics will be present in the phase voltages, but get cancelled in the line–line voltages, and many of the advantages of SVPWM are achieved.

In a balanced three-phase, three-wire system (balanced sources, loads, and line impedances) as shown in Fig. 2.28, the cycle-by-cycle average neutral voltage $\bar{v}_{nN}$ with respect to the DC link negative rail can be derived using $\bar{i}_A(t) + \bar{i}_B(t) + \bar{i}_C(t) = 0$, and is given in Eq. (2.36).

$$\bar{v}_{nN}(t) = \frac{V_d}{2} \qquad (2.36)$$

The neutral voltage is the same as the DC offset in each of the individual poles under the above assumptions. Therefore, under balanced conditions, per-phase analysis can be used to calculate various parameters of interest using only the fundamental frequency components. For example, the phase *a* currents and voltages can be calculated based on phase *a* average equivalent circuit, with the pole-neutral voltage $\bar{v}_{An}$ given by Eq. (2.37). The other phase quantities at the fundamental frequency will have the same magnitude and phase-shift of 120° and 240°, respectively. Example 2.6 illustrates this concept.

$$\bar{v}_{An}(t) = \frac{V_d}{2} v_{cA}(t) \tag{2.37}$$

In a three-wire system, i.e., without a neutral connection, the instantaneous neutral voltage, v_{nN} can switch among V_d, $V_d/2$, and 0. In some applications, a four-wire converter topology, with the fourth leg comprised of two capacitors or two switches (similar to a power pole) and the mid-point connected to the neutral, is used. Under unbalanced conditions there can be significant neutral currents in these applications.

Example 2.6 Consider the three-pole converter shown in Fig. 2.28 connected to a 480 V(L–L), 60 Hz, three-phase grid as shown through an inductive filter with L = 1 mH and R = 0.05 Ω. The DC link voltage is 1000 V. It is desired to inject 50 kW power into the grid at a power factor of 0.9.

(a) Calculate the control voltages required for each of the three poles.
(b) Calculate the CCA current drawn by a three-pole converter from the DC link.
(c) Simulate the above converter and show waveforms of the pole currents and the DC link currents.

(a) We can use per-phase phasor analysis taking advantage of the balanced system.

$$V_{an} = \frac{\sqrt{2}\,480}{\sqrt{3}} = 391.92\,\angle 0°\ \text{V}$$

$$I_a = \frac{2P_o/3}{0.9|V_{an}|} \angle \cos^{-1}(0.9) = 94.5\angle -25.84°\ \text{A}$$

The required pole A voltage with respect to the neutral is given by

$$V_{An} = V_{an} + V_L + V_R$$
$$= V_{an} + j\omega L I_a + R I_a = 412.792\,\angle 4.168°\ \text{V}$$

The required control voltage for pole A in phasor domain is

$$V_{cA} = \frac{V_{An}}{V_d/2} = 0.826\,\angle 4.168°\ \text{V}$$

The required control voltages for each of the poles are,

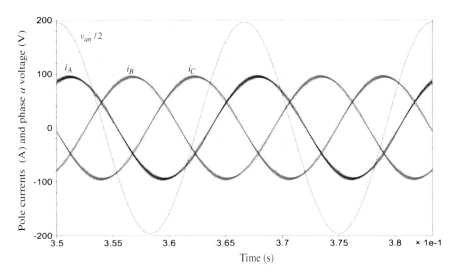

Fig. 2.31 Pole currents of the three-phase converter of Example 2.6

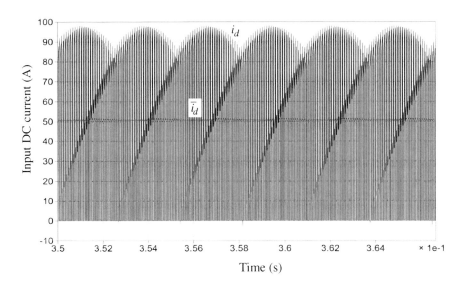

Fig. 2.32 DC link current of the three-phase converter of Example 2.6

$$v_{cA}(t) = 0.826 \cos(377t + 4.168°) \, \text{V}$$
$$v_{cB}(t) = 0.826 \cos(377t - 115.83°) \, \text{V}$$
$$v_{cC}(t) = 0.826 \cos(377t + 124.168°) \, \text{V}$$

(b) The CCA current drawn from DC link only has a DC component and no low frequency harmonics.

$$\bar{i}_d(t) = \frac{P_o + \text{losses in R}}{V_d} = \frac{50000 + 3\,I_{a,\text{RMS}}^2\,R}{500} = 50.67\,\text{A}$$

(c) The instantaneous line currents, i_A, i_B, and i_C obtained from simulation of the switching model and corresponding to the derived control voltages are shown in Fig. 2.31, along with v_{an} for reference. As seen, the currents are balanced and have a peak value of 94.5 A, and the power factor is 0.9. The instantaneous and CCA DC link current are shown in Fig. 2.32. The instantaneous current follows the envelope of the phase currents, and the average current has only the DC component at 50.7 A.

2.6 Other Converter Topologies and PWM Methods

The discussions in the earlier sections of this chapter focused on power pole-based, two-level converter topologies using triangle-comparison PWM methods. These topologies are widely used in several applications especially related to renewable energy grid integration. However, research and development efforts are constantly being pursued on alternate converter topologies that have the potential for higher conversion efficiency, improved power quality and lower requirements on filter components, improved dynamic performance, higher power density, lower stresses on power semiconductor devices, and/or lower cost.

Multilevel VSC topologies are gaining popularity in grid integration of wind generators and photovoltaics at high power and higher voltage levels, mostly in three-phase applications [5, 6]. Similar to the two-level converters, which are formed by one or more SPDT switches, the multilevel converters are formed by one or more single-pole, multiple-throw (SPMT) switches. The advantages include lower voltage stress on the power devices, and inherently lower distortion (high frequency content) in the line voltages and currents, and hence, reduced filter requirements.

For three-phase applications, space vector PWM (SVPWM) is presently a very popular PWM technique. Unlike in the carrier-based PWM approach where each phase is considered separately and compared individually with a common carrier, in the SVPWM approach the three phases are considered together. For a three-phase system with no zero sequence components, we can define a unique transformation between phase voltages (or other phase quantities) and space vector $\vec{v}_s$ as given in Eq. (2.38).

$$\vec{v}_s(t) = v_a(t)\,e^{j0} + v_b(t)\,e^{j\frac{2\pi}{3}} + v_c(t)\,e^{j\frac{4\pi}{3}} \tag{2.38}$$

The reference voltage space vector to be generated by the power converter is obtained from the controller directly in the space vector domain or by applying Eq. (2.38) to reference phase voltages. Similarly, Eq. (2.38) can be applied to the instantaneous pole output voltages. A two-level, three-phase converter as shown in Fig. 2.28 has three bipositional switches with two possible states for each switch. Therefore, there are only a total of eight ($2^3 = 8$) possible states, resulting in eight possible discrete voltage space vectors. The concept of space vector modulation refers to realizing the continuous reference voltage space vector in an average sense by switching at relatively high frequency among the nearest three of the possible discrete space vectors. The availability of powerful digital signal processors (DSPs) has made the implementation of SVPWM relatively easy. SVPWM offers several advantages over carrier-based PWM in terms of higher AC voltage magnitude for a given DC link voltage, better waveform quality under some operating conditions, and more flexibility. Also, when field-oriented control is employed, as is the case in most wind generators, the voltage reference is obtained directly in the space vector domain, making SVPWM the straightforward approach. For detailed discussions on SVPWM concepts and design, the reader is referred to [4].

References

1. Mohan N (2009) First course on power electronics and drives. MNPERE, St Paul
2. Mohan N (2003) Electric drives: an integrative approach. MNPERE, St Paul
3. Mao X, Ayyanar R, Krishnamurthy HK (2009) Optimal variable switching frequency scheme for reducing switching loss in single-phase inverters based on time-domain ripple analysis. IEEE Trans Power Electron 24:991–100
4. Holmes DG, Lipo TA (2003) Pulse width modulation for power converters: Principles and practice. IEEE press series on power engineering. Wiley, New York
5. Rodriguez J, Lai JS, Peng FZ (2002) Multilevel inverters: a survey of topologies, controls, and applications. IEEE Trans Ind Electron 49:724–738
6. Malinowski M, Gopakumar K, Rodriguez J, Pérez MA (2010) A survey on cascaded multilevel inverters. IEEE Trans Ind Electron 57:2197–2206

Chapter 3
Power Converter Topologies for Grid Interface of Wind Energy

This chapter focuses on the power electronic converter topologies used in grid integration of wind generators. The need for variable speed operation in order to capture the maximum possible energy at different wind velocities is discussed first. Based on this requirement, as well as to support emerging requirements on grid support features, a majority of the new wind generator designs employ the doubly fed induction generators with partial rated power converters or permanent magnet synchronous generators with fully rated power converters. Hence, the focus of this chapter is specifically on the power converter topologies for these two types of wind generators. The circuit configurations, basic operation and overview of control structure, and some relevant design details are presented. The capabilities of each of these two types of wind generators to support existing and emerging standards on grid interconnection are discussed

3.1 Variable Speed Operation and Grid Support Requirements

Chapter 1 discussed the maximum power that can be generated as a function of wind velocity. In order to generate this maximum power at any given wind velocity, the wind turbine needs to rotate at a specified optimum speed. The speed-power characteristics of a typical wind turbine at different wind velocities are shown in Fig. 3.1a. The maximum power characteristic curve, which is essentially the locus of the maximum power points for different wind velocities, and the fixed speed characteristic are also plotted in the same figure. The power captured under both the schemes is compared in Fig. 3.1b, highlighting the higher power captured with the variable speed scheme. Since the wind velocity can be less than the rated for significantly long periods, the difference in the energy captured is even more significant. From 3.1a it can be seen that the turbine speed needs to vary from about 0.7 pu (of generator synchronous speed) at the minimum power level to

V. Vittal and R. Ayyanar, *Grid Integration and Dynamic Impact of Wind Energy*,
Power Electronics and Power Systems, DOI: 10.1007/978-1-4419-9323-6_3,
© Springer Science+Business Media New York 2013

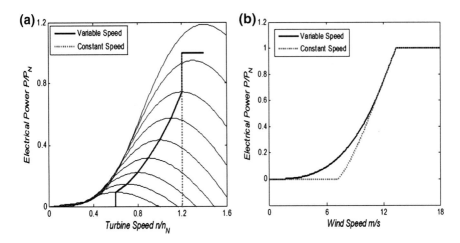

Fig. 3.1 Power capture with constant speed and variable speed wind generators **a** wind speed-power characteristics and **b** comparison of power captured with the two schemes at different wind speeds

about 1.2 pu at the maximum power level, and hence, the wind generator and the power converter topology should be capable of supporting speed changes over a range of 0.5 pu at a minimum (for example, turbine speeds between 10 and 20 rpm for GE 1.5 MW generator). Squirrel cage induction generators and conventional line connected synchronous generators cannot efficiently control the turbine speed in the above range. Hence, DFIGs with a relatively large speed range, and more recently, PMSM generators have become preferred configurations for new designs.

With the power levels of individual wind farms approaching those of conventional power plants, and with the increasing percentage of wind in the overall energy mix for electric power generation, existing and emerging grid interconnection standards require the wind generators to remain online and provide similar response and frequency support during grid faults and other disturbances as provided by conventional generators. Figure 3.2 shows the low voltage ride through (LVRT) requirements specified in [1]. The wind generators are also expected to provide reactive power support and voltage regulation capabilities, similar to or even exceeding those of conventional generators. The requirements on grid support features will influence the selection of wind generator configuration as well as the topology and ratings of the power converter.

3.2 Power Converters in Doubly Fed Induction Generator

The configuration of a doubly fed induction generator is shown in Fig. 3.3 along with the detailed schematic of the power converters used. As seen, it has two back-to-back, three-phase, voltage source converters sharing a common DC link.

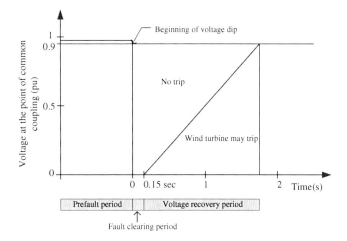

Fig. 3.2 LVRT requirements for wind generators (based on [1])

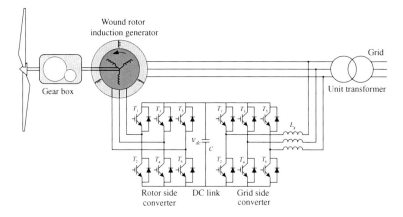

Fig. 3.3 Configuration of DFIG and schematic of the power converters used

The rotor side VSC injects controlled, variable frequency (slip frequency) currents into the rotor of the DFIG through its slip rings. The grid side converter interfaces with the three-phase grid and supports power exchange between the rotor of the DFIG and the grid through the DC link. The grid side converter can also provide reactive power support within its ratings. Both the converters comprise of three power-poles, and the analytical methods explained in detail in Chap. 2 for power poles can be readily extended for analyzing these converters. Both VSCs are capable of supporting bidirectional power flow. This capability is utilized to operate the DFIG in both the supersynchronous mode where power flows from the rotor to the grid, and the subsynchronous mode where the power flows from the grid to the rotor.

3.2.1 Control Functions of Different Stages

The dynamic model of the DFIG and controller design for the two converter stages are discussed in detail in Chap. 4. This section gives an overview of the control functions of the two converters.

The rotor side converter directly controls the active and reactive power flow from the stator of the DFIG to the grid. This is achieved by controlling the magnitude, frequency, and phase angle of the three-phase currents injected into the rotor by the duty ratio (PWM) control of the voltage source converter. The specific control objectives of the rotor side converter are:

1. Maximum power extraction by controlling the torque/power such that the rotor speed always tracks the optimum given by the tracking characteristic shown in Fig. 3.1.
2. Control of reactive power (decoupled from active power control) exchange between the induction generator and the grid. This can be used for voltage regulation at the generator terminals.
3. More recently, supplementary control to provide frequency support to the grid by changing the active power transfer during large disturbances, using the inherent inertia of the wind turbine.

The grid side converter synthesizes at its AC terminals fixed frequency (grid frequency) three-phase voltages at controlled magnitude and phase (with respect to the grid voltage). The specific control objectives of the grid side converter are:

1. Regulate the DC link voltage by providing a path for the active power transfer (positive or negative) between the rotor side converter and the grid.
2. Provide additional reactive power support to the grid similar to the operation of a STATCOM within the constraints of its MVA rating, which is significantly lower than the rating of the wind generator.

Control of both the converters is typically done in the synchronously rotating grid voltage oriented frame of reference, which makes it easy to perform decoupled control of active and reactive power. The pulse-width modulation is typically done using a space vector modulation approach.

3.2.2 Ratings of the Power Converters

One of the main advantages of the DFIG-based wind generator is that the power converters need to be rated only for a fraction of the power rating of the wind generator. For a given range of required speed variation, the ability to operate both in sub- and supersynchronous modes allows the ratings of the converters to be significantly reduced. For example, if the DFIG operates only in the supersynchronous mode, the speed variation would be roughly from 1 to 1.5 pu (of the synchronous speed), and the required rating for both the rotor side and grid side converters will be 0.5 pu of the rated power of the wind generator. This peak power transfer through the converters occurs at high wind velocities and rotor speed at 1.5 pu. However, with both super- and subsynchronous modes, the speed variation would be roughly from 0.7 to 1.2 pu, and the required power rating for each of the converters will be only 0.2 pu of the rated power of the wind generator. The peak power occurs at high wind velocity and with rotor speed at 1.2 pu. At low wind velocities, the power flow is in the reverse direction, and maximum power transfer in this mode occurring at rotor speed of 0.7 pu is less than that at 1.2 pu, since the power at low wind velocities is significantly lower. Hence, in DFIG systems, the rotor side as well as grid side converters are typically rated only for about 20–25 % of the rated power of the system.

The switches are typically implemented with insulated gate bipolar transistors (IGBTs) and anti-parallel diodes. The nominal voltage magnitude at the stator terminals can be between 575 and 690 V L–L for most wind generators above 1 MW rating. There is a move toward higher voltage levels as the power levels increase. The voltage at the stator is stepped up by a unit transformer to 34.5 kV for interconnection to the subtransmission system. The DC link voltage needs to be at least higher than the peak value of the line–line stator voltage for space vector PWM and higher than $\sqrt{3}/2$ times the peak line–line voltage for carrier-based PWM, leading to typical values of DC link voltage in the range of 1200–1500 V. Hence, the voltage ratings of the IGBTs are required to be above 2 kV. The switching frequency at the multi-MW power levels are typically a few kHz. The rotor voltage magnitude varies with the operating slip, and the maximum rotor voltage (at the highest slip frequency) can be in the range of 200–300 V L–L. The turns ratio between the stator and rotor windings is chosen to optimally use voltage and current ratings of commercially available power devices.

3.2.3 Protection During Grid Faults

The wind generators are required by interconnection standards to remain connected to the grid during grid faults and other disturbances when the terminal voltage can be very low. The LVRT specifications are given in Fig. 3.2 [1]. For the DFIG configuration, since the stator is directly connected to the grid (through a step-up transformer), the stator currents during grid faults can be large. This results in large induced rotor currents that also circulate through the PWM converters. The fault current magnitude far exceeds the current ratings of the devices of the converters, and may lead to overvoltage at the DC link. The combination of pitch control (slow) to reduce the wind power output, and PWM control of rotor side converter with its limited voltage capability is usually not sufficient to limit the fault current magnitude below the ratings.

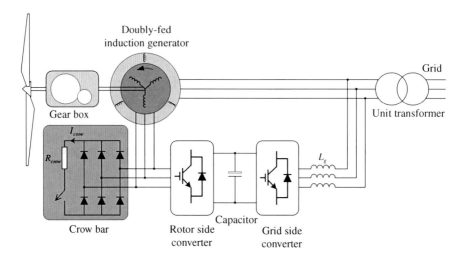

Fig. 3.4 Active crow-bar circuit to bypass and control fault currents

In order to avoid large fault currents through the converters, DFIG-based wind generators employ a protection scheme called the crow-bar circuit. The passive crow-bar circuits essentially involve shorting the rotor windings during fault conditions by external switches, usually with a series resistance, such that the fault currents are bypassed away from the rotor side converter. The DFIG under this condition essentially has the characteristics of a squirrel cage machine with a large rotor resistance. Hence, the generator draws significant reactive power, which may further exacerbate the fault conditions. Providing reactive power support through the grid side converter to partially offset the above drawback is an option. Active crow-bar circuits such as the one shown in Fig. 3.4 are also beginning to be used to bypass the fault currents from the rotor side converter while still providing limited control during grid fault conditions [2, 3]. Here, the effective series resistance to the rotor windings can be controlled by the control of the IGBT in the crow-bar circuit. Active crow-bar circuits are effective in reducing the reactive power drawn by the DFIG during faults and smooth transitions to normal operation when the fault is cleared and the rotor side converter resumes operation.

3.3 Power Converters for Type 4 Wind Generators

Type 4 generators are emerging as an attractive solution to address the increasingly challenging grid interconnection requirements and standards. They use fully rated power converters with permanent magnet synchronous generators. It is also possible to use induction generators or wound-rotor, DC-excited synchronous generators with the fully rated power converters in the same configuration.

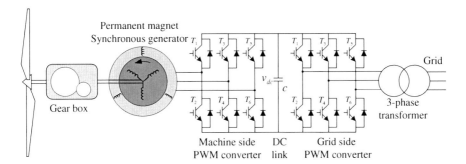

Fig. 3.5 Configuration of a permanent magnet synchronous machine-based wind generator and the schematic of the power converters used

The configuration of a permanent magnet synchronous machine (PMSM)-based wind generator along with the schematic of the power converters is shown in Fig. 3.5. The rotor of the generator has permanent magnets embedded, which produce the air gap magnetic field. This avoids the need for slip rings and brushes, leading to lower maintenance requirement. Permanent magnet machines also tend to have higher efficiency since the losses in the rotor windings are eliminated. It is possible to directly couple the wind turbine to the rotor of the machine without a gear, since the stator is fully decoupled from the fixed-frequency grid. However, since the size, weight, and losses of a machine depend more on the rated torque than on the rated power, it is usually a better tradeoff in terms of size, cost, and efficiency to use a step-up gear to run the rotor at higher speed and lower torque for a given power rating of the generator.

The stator of the PMSM is connected to the machine side converter (MSC). In Fig. 3.5 the MSC is shown as a fully controllable, four-quadrant, three-phase voltage source PWM converter. However, since the power flow is always from the stator to the grid, other simpler topologies such as a diode bridge rectifier followed by a PWM DC–DC converter has also been employed. When a PWM voltage source converter is used, it generates variable frequency three-phase voltages at controlled magnitude at the AC terminals. The grid side converter (GSC) is most often a four-quadrant, voltage source converter as shown in Fig. 3.5, and shares the DC link with the MSC. It generates three-phase voltages at the grid frequency, but controls the magnitude and phase to control power flow.

A key difference between the DFIG and PMSM-based wind generators is that the two converters of PMSM-based generator are each rated for the full power rating of the wind generator. Though it is certainly more expensive, the fully rated converters enable the Type 4 wind generators to provide grid support features that are not possible with DFIG-based wind generators.

The functions of the machine-side converter are:

1. Maximum power tracking under varying wind speeds by controlling the frequency, magnitude, and phase of the three-phase voltages applied at the stator terminals.

2. The power factor is controlled to be unity in order to get the maximum active power for a given MVA rating of the converter.
3. MSC can be controlled to provide inertial and frequency support by utilizing the inherent inertia of the wind turbine during grid contingencies. For example, when the grid frequency drops during a transient, a supplementary control loop can override maximum power tracking, reduce the turbine speed, and release the stored kinetic energy to provide grid support.

The functions of the grid side converter are:

1. Transfer the power from the machine through the MSC to the grid.
2. Regulate the DC link voltage.
3. Provide grid support features as discussed in Sect. 3.3.1. (including zero voltage ride through (ZVRT), support for fault recovery, reactive power support, and voltage regulation).

3.3.1 Performance Under Grid Faults and Other Grid Support Features

The drawbacks of the DFIG include large currents in the machine and converters during grid faults, and the resultant need for crow-bar circuits. In PMSM type wind generators, the machine is fully decoupled from the grid through two voltage source converters. Hence, it is easier to keep the wind generator connected to the grid even for deep voltage sags down to zero volts. The GSC can control the converter output voltage down to zero to limit the fault currents. Furthermore, it can control the current injected into the grid to have any arbitrary phase relationship. This feature together with the capability to continue to source limited active power into the grid even during faults can be used to significantly support fault recovery. Also, as mentioned in the previous section, the MSC can provide inertial and frequency support by utilizing the inherent inertia of the wind turbines.

Under normal grid conditions, the GSC can function as a STATCOM to provide reactive power support to the grid. Though this capability is similar to that of the GSC in DFIG, here the MVAr capability is significantly higher due to the fully rated GSC as opposed to the partial rating in DFIG. The fully rated GSC is also advantageous in interfacing battery storage for mitigating fast variations in wind speed, and providing grid controllable ramp rates.

3.4 Other Emerging Power Converter Topologies

As the power levels of individual wind turbines increase, there is a move toward utilizing higher operating voltages in the range of 3.3 kV and higher, similar to those employed for medium voltage motor drives. The DC link voltage, therefore,

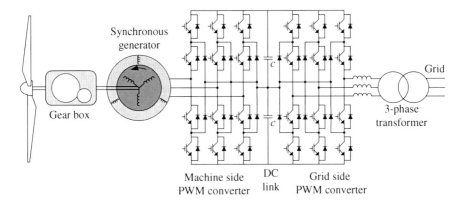

Fig. 3.6 Schematic of a double, three-level PWM converter used in a PMSM-based wind generator

also increases correspondingly and makes it challenging to obtain power semi-conductor devices of suitable voltage rating and high frequency characteristics. Multi-level converters employing more number of switches but each with significantly lower voltage stress are emerging as an attractive topology for wind generators at higher power levels [4]. Figure 3.6 shows the schematic of a three-level PWM converter-based PMSM wind generator. A similar configuration can be used for the DFIG-based wind generators as well. There are four switches per leg, but each rated for only half the DC link voltage. The voltages across the two DC link capacitors are dynamically regulated to be equal by the control of the two PWM converters. A main advantage of the multilevel converter is that each of the pole output voltages has multiple switching levels. For example, in a three level converter shown in Fig. 3.6, the voltage levels are V_{dc}, $V_{dc}/2$, and 0. This results in significantly lower switching frequency content in the line–line currents, and hence, the filter needed to limit the high frequency content is significantly smaller compared to two-level converters discussed earlier for the same switching frequency. The advantages, both in terms of switch voltage stress and power quality, become more significant as the levels increase, but at the expense of more number of switches and higher control complexity.

Other emerging power conversion architectures include DC distribution within a wind farm with each wind generator producing a DC output. The power from each generator is distributed through a DC bus, which is converted by a single high power inverter and connected to the grid at the wind farm site, or alternately as done in off-shore applications, transmitted through high voltage DC cables close to point of use before conversion to AC. The scheme potentially avoids multiple stages of power conversion and results in efficiency improvements. Also, the DC bus offers a convenient and efficient node for adding energy storage.

References

1. The Technical Basis for the New WECC Voltage Ride-Through (VRT) Standard. White paper by the Wind Generation Task Force1 (WGTF) of WECC, June 2007
2. Peng Z, Yikang H (2007) Control strategy of an active crowbar for DFIG based wind turbine under grid voltage dips. In: Proceeding international conference on electrical machines and systems, Seoul, Korea, 8–11 Oct 2007
3. Jin C, Wang P (2010) Enhancement of low voltage ride through capability for wind turbine driven DFIG with active crowbar and battery energy storage system. In: Proceeding IEEE power and energy society general meeting
4. Ma K, Blaabjerg F (2011) Multilevel converters for 10 MW Wind Turbines. In: Proceeding 14th European conference on power electronics and applications (EPE)

Chapter 4
Control of Wind Generators

A wind generator has multiple control objectives including maximum power extraction from wind for a given wind velocity within the constraints of its various ratings, control of reactive power exchange with the grid and possibly voltage regulation, low voltage ride through (LVRT) and fault ride through (FRT), and internal control objectives such as regulation of the DC link voltage in the case of DFIG and generators with fully rated converters (FRC). We focus on the control of DFIG-based wind energy systems in this book since it is the most prevalent technology at present. The model of a DFIG machine was presented in Chap. 1 using the d-q reference frame. Analysis using the d-q frame of reference naturally lends well to the design of decoupled control of active power (or torque) and reactive power, which is the main focus of this chapter. By controlling the rotor-injected currents by the rotor side converter, the stator currents and active and reactive power can be controlled. The grid side converter facilitates the power exchange through the rotor converter, and also provides additional reactive power support.

This chapter begins with a description of steady-state operation including analysis of the required rotor currents and voltages under different wind velocities leading to subsynchronous and supersynchronous modes, in order to gain a good understanding of the control mechanisms. This then leads to a dynamic analysis of DFIG systems, descriptions of transfer functions of interest, and finally controller design for the two PWM converters. Many of the concepts are illustrated through numerical simulation of an example DFIG system with parameters corresponding to a 1.5 MW wind generator.

V. Vittal and R. Ayyanar, *Grid Integration and Dynamic Impact of Wind Energy*,   65
Power Electronics and Power Systems, DOI: 10.1007/978-1-4419-9323-6_4,
© Springer Science+Business Media New York 2013

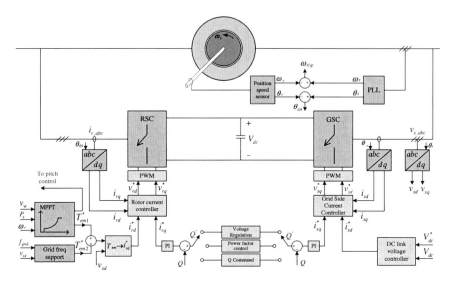

Fig. 4.1 An overview of the control of a DFIG wind generator system

4.1 Overview of Control of DFIG-Based Wind Generator System

The configuration of a DFIG-based wind generator system with dual PWM converters—the rotor side converter (RSC) and grid side converter (GSC) was discussed in Chap. 3 (refer Fig. 3.3). The various control objectives for the DFIG wind system and an overview of their implementation are illustrated in Fig. 4.1, and discussed further in this section [1, 2]. Later sections elaborate on the details of the individual control blocks of Fig. 4.1.

Rotor side converter control:

- The main control objective is to capture the maximum possible power from the wind under a given operating condition (mainly wind speed). The electromagnetic torque of DFIG is controlled such that the rotor always operates at the optimal speed given by maximum power tracking characteristics. The torque reference is obtained by a speed control loop. The rotor d-axis current controller ensures the torque is maintained at the reference value. In grid voltage-oriented control described in subsequent sections, the angle required for d-q transformation is obtained from the grid voltage using a phase-locked loop (PLL). The torque command from maximum power tracking control loop may be modified during system contingencies to provide frequency support to the grid by utilizing the stored kinetic energy of the wind turbines.
- In a squirrel cage induction generator of a given design the reactive power absorbed at the stator terminals is a function of the active power, slip and grid

voltage, and cannot be controlled independent of the active power. However, in a DFIG both the active power and the reactive power at the stator can be controlled independently through the control of rotor-injected currents. This can be used to provide voltage regulation or reactive power support to the grid within the ratings of the system. The q-axis rotor current is controlled to control the reactive power.

  Grid side converter control:

- The DC link voltage is regulated by the active power control of the GSC through its d-axis current component. The voltage is regulated with good dynamics such that voltage ratings of the devices in the two PWM converters are not exceeded.
- In addition to the reactive power control of the RSC, the reactive power output of the GSC can be controlled independently through its q-axis current component. Note that i_{sd} and i_{sq} in the GSC context refer to the d-axis and q-axis currents respectively in the grid side PWM converter, and not the stator currents.

4.2 Steady-State Analysis of DFIG with Per-Phase Equivalent Circuit

The per-phase equivalent circuit of a DFIG corresponding to a given slip, s, can be derived from the basic principles of operation of induction machines and transformers. Such an equivalent circuit is quite useful for steady-state analysis with phasors. In order to be consistent with the dynamic models introduced in subsequent sections as well as with models used in the literature, the current directions in the equivalent circuits are defined based on the motor convention.

4.2.1 Development of Per-Phase Equivalent Circuit

Figure 4.2 illustrates the development of a single-phase equivalent circuit, with the rotor impedances referred to the rotor side and the stator quantities referred to the stator side, coupled by the equivalent model of the rotating transformer. R_s and R_r are the physical resistances of the stator and rotor windings, respectively, L_{ls} and L_{lr} are the leakage inductances of the stator and rotor windings, and L_m is the magnetizing inductance referred to the stator side. V_s is the grid voltage applied at the stator and V_r is the rotor voltage applied by RSC, both represented as phasors; I_s and I_r are the stator and rotor current phasors, respectively, as indicated. The core losses are neglected in the model. However, it can be included if required by adding a shunt resistor of suitable value across the magnetizing inductance.

  It is easy to see that the rotating structure scales the frequency by a factor equal to the slip s at the rotor since the magnetic field quantities seen by the rotor windings rotate at the slip speed. The frequency of the stator side quantities (V_s, I_s, I_m) is ω_s rad/s and the frequency of the rotor side quantities (V_r, I_r) is $s\omega_s = \omega_{slip}$ rad/s. The voltages are also scaled by a factor equal to s. This can be explained from the basic principle that induced voltages are proportional to the relative motion between the windings and the field. The stator is stationary, and therefore, the relative velocity with any field quantity (which rotates at ω_s) is ω_s, while the rotor rotates at a velocity ω_r and its relative velocity with respect to any field quantity is $\omega_s - \omega_r = s\omega_s$. Therefore, for example, the magnitudes of stator side voltages when referred to the rotor side are multiplied by s. In addition, if the physical turns ratio between the stator and rotor windings is $N_r/N_s = n$, then the rotor voltage is scaled by an additional factor of n. So, the primary voltages when referred to the rotor side are multiplied by the factor ns as indicated in Fig. 4.2b.

Unlike the voltage magnitudes, the current magnitudes are not scaled by s when moving across the rotating transformer. This is because the currents drawn from the stator of an induction machine to cancel the *MMF* produced by the rotor currents do not depend on the rotation or relative velocity of the individual windings. Only the ampere-turns in the stator windings are required to be equal to the ampere-turns in the rotor windings, with the assumption of the leakage inductance being negligible compared to the magnetizing inductance. Therefore, for example, when referring any current from the stator side to the rotor side, the current magnitude is multiplied by the physical turns ratio n alone. This is illustrated in Fig. 4.2b with the controlled current source at the stator side, which is equal to $n\,I_r$. It is important to remember that the frequency of the current is scaled by s. Since only the voltage is scaled by s and not the current, the impedances are scaled by s when transferring from stator to rotor (and not by s^2 as is the case for physical turns ratio), and $1/s$ when transferring from rotor to stator.

In a transformer with stationary primary and secondary windings the power processed by the primary and secondary windings are equal, which is consistent with the scaling of the voltage by the turns ratio n and current by $1/n$ when moving from the primary to the secondary. However, in an induction machine the power processed by the stator and the rotor windings are not equal as can be inferred from the scaling of only the voltage by s and not the current, and the value of power (active and reactive) processed by the two controlled sources not being equal. The power processed by the stator winding at the air gap P_g (which does not include stator resistive loss) and the power processed by the rotor winding at the air gap P_r are related by (4.1) as seen from Fig. 4.2b.

$$P_r = s P_g \qquad (4.1)$$

The difference in the active power processed between the two windings is the mechanical power, absorbed or sourced, depending on operation as a motor or generator, respectively, and is given in (4.2) and illustrated in Fig. 4.2b (motor convention). This is the power required to maintain the rotor at a velocity of ω_r at a torque of T_{em} (again positive or negative depending on operating mode).

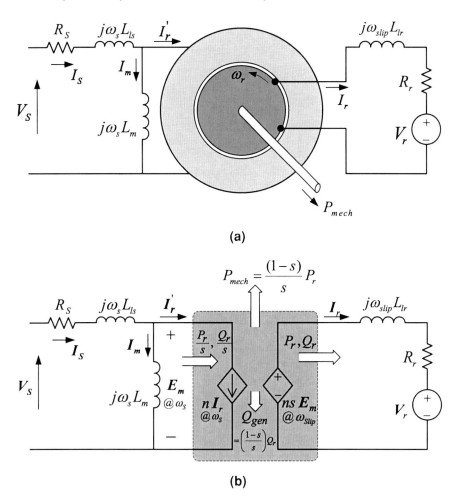

Fig. 4.2 Development of a per-phase equivalent circuit of a DFIG

$$P_g - P_r = (1-s)P_g = \left(\frac{1-s}{s}\right)P_r = P_{mech} = \omega_r T_{em} \qquad (4.2)$$

A similar expression can be derived for the reactive power relationship also, as shown in (4.3), with the rotating transformer structure scaling the reactive power by s. This is an important relationship as it shows that the reactive power at the stator can be controlled by a significantly smaller reactive power injection at the rotor. Unlike the active power case, where the difference between P_r and P_g is the mechanical power injected or absorbed, in the case of reactive power, the difference between Q_r and Q_g is generated by the DFIG.

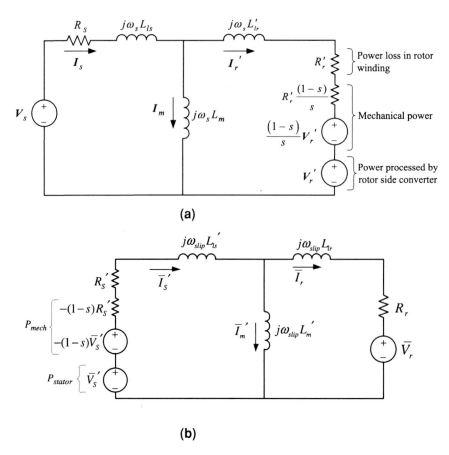

(a)

(b)

Fig. 4.3 Per-phase equivalent circuit of a DFIG at a given slip **a** referred to the stator side and **b** referred to the rotor side

$$Q_r = sQ_g$$

$$\therefore Q_g - Q_r = (1 - s)Q_g = \left(\frac{1-s}{s}\right)Q_r = Q_{gen} \tag{4.3}$$

Based on the discussions above, the per-phase equivalent circuit can be obtained referred to the stator side as shown in Fig. 4.3a or referred to the rotor side as shown in Fig. 4.3b. Though the two equivalent circuits provide identical results, for power system studies the equivalent circuit referred to stator side is perhaps easier to use as it directly gives the current, active power, and reactive power injected to the grid. The equivalent circuit referred to the rotor side has the advantage in that it avoids dividing the rotor resistance and converter voltage by zero when the slip is zero. In both equivalent circuits, the physical turns ratio between stator and rotor windings has been accounted by suitable scaling factors to make the equations less cumbersome. Therefore, in the stator-referred equivalent

Table 4.1 Parameters of the example DFIG machine analyzed

Rated power, P_e	1.5 MW
Rated stator voltage	575 V L–L (RMS)
Grid frequency	60 Hz
Stator resistance, R_s	0.0046 Ω
Rotor resistance, R_r	0.0032 Ω
Stator leakage inductance, L_{ls}	0.0947 mH
Rotor leakage inductance, L_{lr}	0.0842 mH
Magnetizing inductance, L_m	1.526 mH
Number of poles, p	6

circuit of Fig. 4.3a, $V_r^{'} = V_r/n$, $L_{lr}^{'} = \dfrac{L_{lr}}{n^2}$ and $R_r^{'} = \dfrac{R_r}{n^2}$, and in the rotor-referenced equivalent circuit of Fig. 4.3b, $V_s^{'} = n V_s$, $L_{ls}^{'} = n^2 L_{ls}$ and $R_s^{'} = n^2 R_s$ where n is ratio of rotor to stator turns, $\dfrac{N_r}{N_s}$.

Referring to the stator-referred equivalent circuit of Fig. 4.3a, the rotor resistance referred to the stator side $\dfrac{R_r^{'}}{s}$ is shown as the sum of two components $R_r^{'}$ and $R_r^{'} \frac{(1-s)}{s}$. Similarly, the injected voltage $\dfrac{V_r^{'}}{s}$ is shown as the sum of two components $V_r^{'}$ and $V_r^{'} \frac{(1-s)}{s}$. The power loss in the resistor $R_r^{'}$ corresponds to the physical power loss in the rotor winding resistance due to the actual rotor current; the power processed by the voltage source $V_r^{'}$ corresponds to the actual power processed (absorbed or supplied) by the rotor side converter; and the sum of the power in the resistance $R_r^{'} \frac{(1-s)}{s}$ and the voltage source $V_r^{'} \frac{(1-s)}{s}$ corresponds to the actual mechanical power (absorbed or supplied) and is consistent with the expression for mechanical power given in (4.2). Therefore, the expression for the total mechanical power considering all the three phases, and the expression for the total torque are as given in 4.4 and 4.5, respectively.

$$P_{mech} = 3\left|I_r^{'}\right|^2 R_r^{'} \left(\frac{1-s}{s}\right) + 3Re\left[V_r^{'}\frac{(1-s)}{s}I_r^{'*}\right] \tag{4.4}$$

$$T_{em} = \frac{P_{mech}}{\omega_r} = \frac{1}{\omega_r}\left[3\left|I_r^{'}\right|^2 R_r^{'} \left(\frac{1-s}{s}\right) + 3Re\left[V_r^{'}\frac{(1-s)}{s}I_r^{'*}\right]\right] \tag{4.5}$$

4.2.2 Speed-Torque Characteristics at Different Rotor Voltages

The equivalent circuit of Fig. 4.3a can be used to obtain the speed-torque characteristics of a DFIG machine. The speed-torque characteristics corresponding

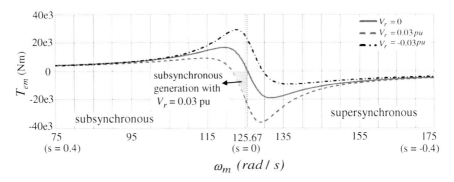

Fig. 4.4 Speed-torque characteristics of a DFIG at different rotor injections

to the parameters given in Table 4.1 are obtained using (4.5) for a range of slip varying from -1 to 1 at three different, fixed values of rotor converter voltage: 0, 0.03, and -0.03 pu (pu of stator voltage). The rotor converter voltages are in phase with the stator voltage in this example, and the turns ratio between stator and rotor windings is assumed to be one.

The rotor current required in (4.5) can be obtained by considering the Thévenin equivalent circuit of Fig. 4.3a looking into the stator side. The resulting expression for the rotor current as a function of the slip is given in (4.6).

$$I'_r(s) = \frac{V_{Th} - \dfrac{V'_r}{s}}{\dfrac{R'_r}{s} + jX'_{lr} + Z_{Th}} \tag{4.6}$$

where, $V_{Th} = \frac{jX_m}{R_s + j(X_{ls} + X_m)} V_s$ and $Z_{Th} = jX_m \| (R_s + jX_{ls})$

Figure 4.4 shows the resulting speed-torque characteristics for a slip range of -0.4 to 0.4, which corresponds to rotor speeds $((1-s)377(2/p))$ in the range of 75–175 rad/s. The curve in the middle with $V_r = 0$ is the same as that for a squirrel cage induction machine, with positive slip corresponding to subsynchronous operation resulting in positive torque and motoring mode, and negative slip corresponding to supersynchronous operation resulting in negative torque and generating mode. The rotor-injected voltage significantly alters the speed-torque characteristics. With $V_r = 0.03$ pu, i.e., with the rotor-injected voltage in phase with the grid voltage, negative torque and generation mode are possible even at subsynchronous speeds as shown in the shaded area of Fig. 4.4. For magnitudes of V_r larger than sV_s the current I'_r is negative resulting in negative torque. Under this condition, power is absorbed from the voltage source at the rotor side. Larger magnitudes of V_r lead to generating mode of operation for a wider range of positive slip.

The upper curve in Fig. 4.4 corresponds to out-of-phase injection, i.e., $V_r = -0.03$ pu, and as seen the resulting changes are opposite to that of in-phase injection. It may be noted that in practice the injected rotor voltage is controlled

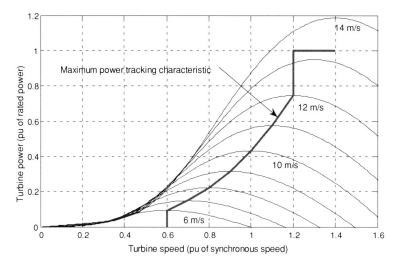

Fig. 4.5 Turbine characteristics and maximum power tracking characteristics

continuously within approximately ± 0.3 pu of grid voltage and over a wide range of phase angles in order to capture maximum power from the wind as the wind velocity changes.

4.2.3 Steady-State Analysis at Various Wind and Rotor Speeds

The equivalent circuits developed in earlier sections can be used to analyze various stator and rotor currents, voltages, and active and reactive power for any chosen operating condition, as a generator or a motor. In this section, we focus on analyzing the operating conditions over the complete range of wind speeds, and therefore, rotor speed variation typical of Type 3 generators. In particular, we will analyze the requirements on the magnitude and phase of the rotor-injected voltages for different wind/rotor speeds to capture maximum power from the wind. We will consider the case where the power injected into the grid is at unity power factor as well as cases corresponding to controlled reactive power injection.

Figure 4.5 shows the turbine characteristics and the maximum power tracking characteristics of an example DFIG-based wind generator for wind speeds ranging from 6 to 14 m/s. The curves roughly correspond to the machine whose parameters are given in Table 4.1. As seen from Fig. 4.5, for any given wind speed, there is a unique rotor speed at which the maximum power occurs. The maximum power tracking characteristic corresponds to operation of the DFIG at this optimum rotor speed and the corresponding maximum power, for a given wind speed.

As seen from (1.7), the power in wind is proportional to the cube of the wind speed. The power coefficient C_p is maintained constant at its optimal value by keeping the tip speed ratio λ constant. Therefore, the rotor speed is roughly proportional to the wind velocity, leading to the approximate relationship between power and rotor speed along the peak tracking characteristic, as given in (4.7).

$$P_{mech} = k_{opt}\, \omega_r^3 \tag{4.7}$$

The constant k_{opt} depends on factors such as the wind velocity at which the rated power is obtained, synchronous (grid) frequency, slip at the maximum power, number of poles, and gear ratio. For a given machine and turbine design, some of the above parameters such as the gear ratio and the slip at which maximum power occurs can be chosen such that the required power rating of the power converter is minimized. This determines the operating range of slip covering positive and negative values. In the analysis below, k_{opt} is chosen such that the slip at the maximum power is -0.2, and is given in (4.8). For the values given in Table 4.1, and an assumed efficiency of 0.93, k_{opt} is calculated to be 0.4703

$$k_{opt} = \frac{P_{\max}/\eta}{\left[\left(1 - s|_{P_{\max}}\right)\omega_s\right]^3} \tag{4.8}$$

where, $s|_{P_{\max}}$ is the desired value of slip at rated power, and η is the efficiency of the DFIG at the rated power.

Considering *unity power factor operation at the stator* terminals initially, the following steps give the required magnitude and phase angle of voltage to be applied by the rotor power converter, the active and reactive power flow through the rotor converter as well as the rotor currents for a wide range of slip variation. Referring to the equivalent circuit shown in Fig. 4.3a,

$$I_s = \frac{P_s/3}{V_s} = \frac{\eta\, k_{opt}\, \omega_r^3}{3(1 - s)}\, \frac{1}{V_s} \tag{4.9}$$

In (4.9), the efficiency of the DFIG over the entire operating range is assumed to be constant in the interest of simple closed-form expressions and ease of analysis. The efficiencies at the stator and rotor sides are also assumed to be equal. For the objectives of the analysis described here, this method gives results that are very close to those from more elaborate, iterative methods to calculate I_s.

$$I_r' = I_s - I_m$$

where, $I_m = \dfrac{E_m}{jX_m}$ and $E_m = V_s - I_s(R_s + jX_{ls})$ $\tag{4.10}$

$$V_r' = s\left(E_m - I_r'\left(\frac{R_r'}{s} + jX_{lr}'\right)\right) \tag{4.11}$$

$$P_{r,conv} = 3\,\mathbf{Re}\left(V_r\, I_r^*\right) \quad \text{and} \quad Q_{r,conv} = 3\,\mathbf{Im}\left(V_r\, I_r^*\right) \tag{4.12}$$

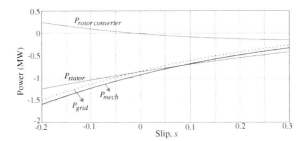

Fig. 4.6 Mechanical, stator, grid, and rotor converter power at different slip, corresponding to maximum power capture from wind at unity power factor

Figure 4.6 shows the plots of mechanical (wind) power, stator power, rotor converter power, and power from grid (motor convention) for various values of slip ranging from -0.2 (maximum power) to 0.3 (minimum power). Though the model is valid for all values of slip, the analysis here focuses on slip values between -0.2 and 0.3, which correspond to the typical range of slip variation in Type 3 generators for varying wind velocities, and covers both subsynchronous and supersynchronous modes. Since motor convention is used, the mechanical, stator, and grid power are negative. The mechanical power obtained from two methods—using (4.4) and (4.7)—is plotted superimposed on each other, and they match quite well. For the rotor converter power, positive values mean power flow is from the rotor to the grid. In the supersynchronous mode of operation between $s = -0.2$ and $s = 0$, the rotor converter power is positive, and hence, power is absorbed from the rotor and fed into the grid through the RSC and GSC. For the subsynchronous mode of operation between $s = 0$ and $s = 0.3$, the rotor converter power is negative, and hence, power is absorbed from the grid through GSC and fed into the rotor. It may be noted from Fig. 4.6 that due to the small loss in rotor winding, the transition from positive to negative power in the rotor converter occurs at $s = -0.013$ instead of exactly at $s = 0$.

The reactive power from stator Q_{stator}, the reactive power absorbed by the rotor converter $Q_{rotor\,converter}$, and the reactive power consumed in the magnetizing and leakage inductances of the DFIG, Q_L, along with the internally generated reactive power by the DFIG, Q_{gen}, are shown in Fig. 4.7 corresponding to the unity power factor operation at the stator terminals. The reactive power Q_L defined in (4.13) is dominated by the reactive power consumed by the magnetizing inductance L_m, and hence, does not change significantly over the operating range as seen from Fig. 4.7.

$$Q_L = Q_{L_m} + Q_{L_{ls}} + Q_{L_{lr}} = 3\left[|\mathbf{I}_m|^2\omega_s L_m + |\mathbf{I}_s|^2\omega_s L_{ls} + |\mathbf{I}_r|^2\omega_{slip}L_{lr}\right] \quad (4.13)$$

The internally generated reactive power Q_{gen} equals the sum of all the reactive power consumed by the magnetizing and leakage inductances, the reactive power absorbed by the grid at the stator terminals (zero for unity power factor operation), and by the rotor converter.

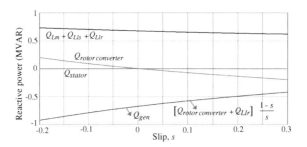

Fig. 4.7 Reactive power balance in a DFIG at unity power factor operation at the stator

$$-Q_{gen} = Q_L + Q_{rotor\,converter} - Q_{stator} \qquad (4.14)$$

From the discussions of Sect. 4.2.1, Q_{gen} is given by the expressions of (4.15).

$$
\begin{aligned}
Q_{gen} &= \left(\frac{1-s}{s}\right)Q_{r,gap} = \left(\frac{1-s}{s}\right)(Q_{rotor\,converter} + Q_{L_{lr}}) \\
&= (1-s)Q_{s,gap} = (1-s)\left[Q_{stator} - Q_{L_m} - Q_{L_{ls}}\right]
\end{aligned}
\qquad (4.15)
$$

Figure 4.7 shows Q_{gen} obtained independently from (4.14) and (4.15) super-imposed, and the two plots match quite well. Also, it may be noted from Fig. 4.7 that the rotor converter processes only a fraction of the total reactive power requirement. For unity power factor operation at the stator, it absorbs reactive power during negative slip, and sources reactive power during positive slip. Unlike the case for active power, where the power absorbed by the rotor converter is equal to the power supplied by the grid side converter (neglecting losses in the converter), the reactive power in the rotor side converter and grid side converter can be totally independent of each other.

The maximum power tracking as well as reactive power control is achieved by applying appropriate rotor voltages by the RSC. The required magnitude (scaled by the stator-rotor windings turns ratio) and phase angle of the rotor converter voltage obtained from (4.11) for the case of unity power factor operation and maximum power tracking are shown in Fig. 4.8a, b, respectively. The frequency of the converter voltage and current equals the operating slip times the grid frequency. The magnitude of sV_s is also shown in the magnitude plot for comparison. As seen, the required magnitude of rotor converter voltage closely follows sV_s, since a major part of the converter voltage goes to cancel the induced voltage approximately equal to sV_s and the difference corresponds to the voltage drops across L_{lr} and R_r due to the required rotor current. Hence, the magnitude of V_r is highest at the maximum operating slip and is close to zero (with frequency also close to zero) at zero slip or synchronous operation. The phase angle in Fig. 4.8b is shown with respect to the stator voltage. The converter voltage is roughly in phase (phase angle close to zero) during subsynchronous mode and out-of-phase (roughly $180°$) during supersynchronous mode. These are consistent with the requirement for the currents I_s and I_r' in Fig. 4.3a to be negative for generator mode operation.

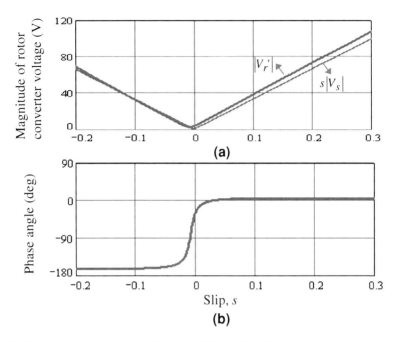

Fig. 4.8 Required rotor converter voltage at different slips for unity power factor operation
a magnitude and **b** phase angle with respect to stator voltage

Fig. 4.9 Polar plot of rotor converter voltage at different slips corresponding to unity power factor operation

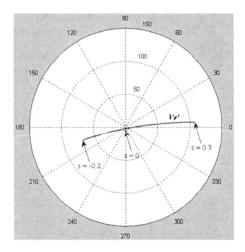

The magnitude and phase angle of rotor converter voltage can also be viewed together on the same polar plot as shown in Fig. 4.9, where each point corresponds to the phasor of the rotor converter voltage at a given operating slip.

Figure 4.10a shows the magnitudes of stator and rotor currents along with the stator magnetizing current, and Fig. 4.10b shows the phase angles of the same currents with respect to the stator voltage. It may be noted that the rotor current is

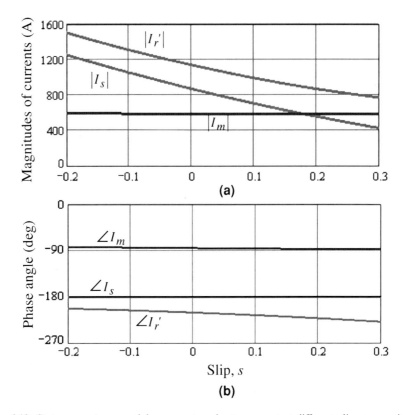

Fig. 4.10 Stator current, magnetizing current, and rotor current at different slips **a** magnitude and **b** phase angle with respect to stator voltage

at slip frequency and the stator and magnetizing currents are at grid frequency. The stator current plots confirm unity power factor operation, and its magnitude corresponds to the power injected into the grid from the stator side. The magnetizing current remains almost constant throughout the range of slip variation. The rotor current is larger than the stator current since it supports the magnetizing current as well. The same current flows through the rotor side converter if the stator-rotor windings turns ratio is one. Typically, the turns ratio is selected to be a higher value (higher number of rotor turns) to reduce the current rating with a higher voltage rating for the rotor converter.

A phasor diagram with plots of the various current and voltage phasors under different operating conditions helps in a clear understanding of the operation of DFIG in subsynchronous and supersynchronous modes. Figure 4.11a shows the phasor diagram based on the equivalent circuit of Fig. 4.3a and corresponding to unity power factor operation at a slip of −0.2, which corresponds to supersynchronous operation at rated power. Figure 4.11b shows a similar phasor diagram for operation at a slip of +0.2 with a correspondingly smaller power

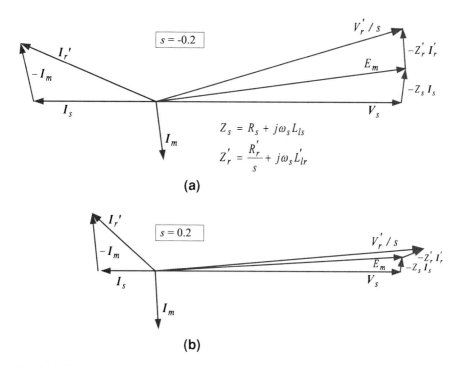

Fig. 4.11 Phasor diagram of DFIG operation at unity power factor at stator terminals **a** supersynchronous mode at slip of -0.2 and **b** subsynchronous mode at slip of 0.2

captured from wind. Note that the quantity $\dfrac{V_r'}{s}$ remains positive for both sub- and supersynchronous modes of operations due to the scaling by s, but the rotor converter voltage V_r changes polarity with respect to the stator voltage based on the operating mode, as discussed earlier.

Similar to the analysis carried out above for unity power factor operation at the stator, analysis under any other power factor or under any given reactive power command within the ratings of the converter and DFIG machine can also be done using the equivalent circuits of Fig. 4.3. The analysis here corresponds to the reactive power at the stator, and does not consider the reactive power from the grid side converter. For an arbitrary power factor angle at the stator, (4.9) can be modified as in (4.16); the rest of the analysis, to determine the required rotor-injected voltage, is similar to that of unity power factor operation

$$I_s = \frac{P_s/3}{V_s\cos(\theta_{I_s})} \angle \pm \theta_{I_s} = \frac{\eta k_{opt}\, \omega_r^3}{3(1-s)}\, \frac{1}{V_s\cos(\theta_{I_s})} \angle \pm \theta_{I_s} \qquad (4.16)$$

where, $\cos(\theta_{I_s})$ is the required power factor.

The operation under 0.9 leading power factor, i.e., the DFIG system sourcing reactive power to the grid, with maximum active power tracking is analyzed here.

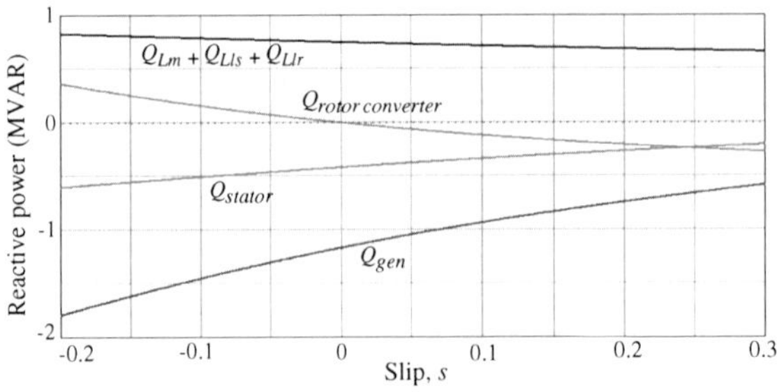

Fig. 4.12 Stator reactive
power, rotor converter
reactive power, and internally
generated reactive power at
different slips corresponding
to maximum power capture
from wind at a leading power
factor of 0.9

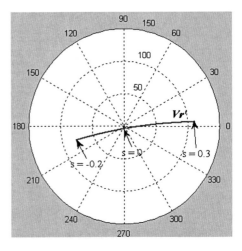

Fig. 4.13 Polar plot of rotor
converter voltage at different
slips corresponding to 0.9
leading power factor

It is assumed that the machine and converters are rated to operate at full power
under ± 0.9 power factor. The plots of various active powers are close to those
shown for the case of the unity power factor operation in Fig. 4.6 except for small
differences in the winding resistance losses due to the difference in the reactive
power components of the rotor and stator currents between the two operating
conditions. The plots of stator reactive power, rotor converter reactive power, and
the internally generated reactive power Q_{gen} are quite different and are shown in
Fig. 4.12 for the leading power factor operation. Q_{gen} is plotted using each of the
three expressions, shown in (4.14) and (4.15), and the three match exactly. Similar
to the unity power factor operation, the reactive power through the rotor converter,
in this case, is also only a small fraction of the total reactive power processed.

    Figure 4.13 shows the polar plot of the required rotor converter voltage to
achieve the above power factor requirement and maximum power tracking at
different values of slip. Since the required converter voltage is dominated by the
component $s V_s$, the plot looks similar to that of the unity power factor case, but the
voltage drops across the rotor impedances and therefore, the rotor currents are
significantly different between the two operating conditions, as seen from the

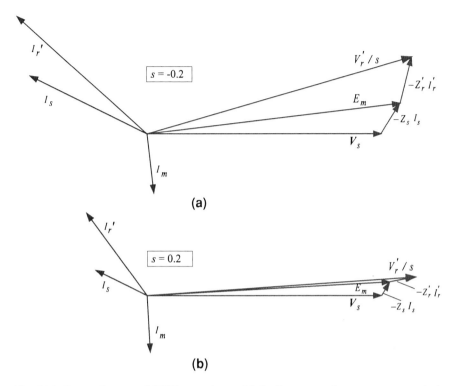

Fig. 4.14 Phasor diagram of DFIG operation at 0.9 leading power factor at stator terminals **a** supersynchronous mode at slip of -0.2 and **b** subsynchronous mode at slip of 0.2

phasor diagram of Fig. 4.14 at slip values of -0.2 and 0.2 corresponding to the leading power factor operation.

4.3 Dynamic Analysis of DFIG and Design of Controllers

The various blocks shown in the overall control block diagram of Fig. 4.1 are analyzed, and suitable control design methods for each of the control loops based on the grid voltage orientation scheme are described in this section.

4.3.1 Torque or Active Power and Reactive Power References

The torque or active power reference for the DFIG control is mainly obtained from the turbine control. The objective of the turbine control is to draw maximum

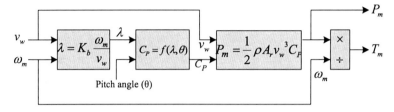

Fig. 4.15 Wind power model

possible power from the wind while operating the turbine within the allowable speed range, and the DFIG and power converters within their current and voltage ratings. When the power available from wind is above the rated power of the generator, pitch control is activated to limit the power and turbine speed. Below the rated power, electrical power or torque is controlled in one of several ways to achieve maximum power tracking as described below.

The wind power model based on the wind velocity and turbine angular speed is shown in Fig. 4.15. The mechanical power output, given in (1.7) depends on the power coefficient C_p, which is a function of the tip speed ratio λ for a given pitch angle θ. The turbine or the rotor speed is continuously adjusted as the wind velocity changes in order to keep C_p at the optimal value, and the output power at its maximum value. The maximum power tracking characteristic like the one shown in Fig. 4.5 for a given turbine design can be obtained from the wind power model, and can be used in the design of turbine control.

Following the maximum power tracking characteristic, for a given wind speed, the maximum power and the rotor speed are fixed. Hence, one of the control methods is to give the power command as a function of the rotor speed. For example, consider the operating point A in Fig. 4.16 at a wind speed of 10 m/s. If the wind speed increases to 12 m/s, the power from wind increases as indicated at point B while the electric power output has not yet increased. This accelerates the rotor increasing the rotor speed, and therefore, the electric power command. The operating point moves along the characteristic corresponding to wind speed of 12 m/s as shown in the figure till the point C is reached, where the electric power has increased to the maximum possible value for wind speed of 12 m/s. A similar analysis can be done to confirm the maximum power tracking for decreasing wind speed as well. An approximation of the maximum power tracking characteristic is given in (4.7) and (4.8).

An equivalent method for turbine control is to obtain the rotor speed reference from the maximum power tracking characteristic as a function of the measured electric power [3], and to regulate the rotor speed to this reference value through closed loop control. The block diagram for the rotor speed loop (along with other control loops) is shown in Fig. 4.17.

The design of the speed controller $G_\omega(s)$ depends on the model assumed for the turbine drive train system—lumped model or two-mass model. The two-mass model of the drive train is shown in Fig. 4.18 from which the transfer function

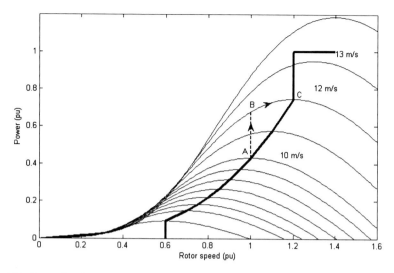

Fig. 4.16 Example of turbine control

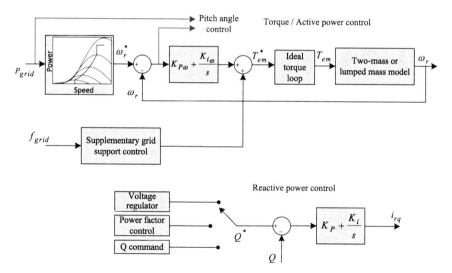

Fig. 4.17 Generation of torque and reactive power references

$\dfrac{\omega_r(s)}{T_{em}(s)}$ needed for the design of the speed loop controller can be derived [4, 5] as given in (4.17).

$$\frac{\omega_r(s)}{T_{em}(s)} = \frac{1}{(J_t + J_g)s} \frac{J_t s^2 + D_{tg}s + K_{tg}}{\frac{J_t J_g}{J_t + J_g} s^2 + D_{tg}s + K_{tg}} \tag{4.17}$$

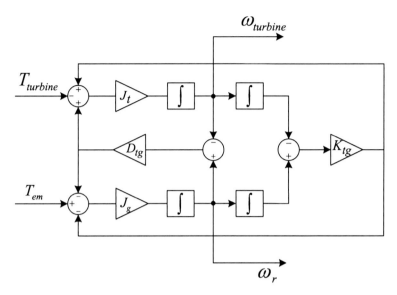

Fig. 4.18 Block diagram of a two-mass drive train system

If the lumped-mass model is used in simplified simulations, the transfer function is reduced to (4.18).

$$\frac{\omega_r(s)}{T_{em}(s)} = \frac{1}{(J_t + J_g)s} \tag{4.18}$$

The output of the speed controller is the torque reference T_{em}^*, which is used to derive the references for rotor d-axis current. For the purpose of speed loop design, the actual torque T_{em} can be assumed to ideally track T_{em}^*, and therefore, the speed controller only needs to compensate for the transfer function $\frac{\omega_r(s)}{T_{em}(s)}$ of (4.17) or (4.18).

The reactive power reference is obtained from a voltage regulation loop that regulates the voltage at the point of interconnection to the grid or from a set power factor command or from a direct system-level command for reactive power support as shown in Fig. 4.17. The error in reactive power is used to control the q-axis rotor current.

4.3.2 Grid Voltage Orientation

Control of the rotor converter is done in the synchronously rotating reference frame with the d- and q- axes rotating at the grid frequency ω_s and with the q-axis

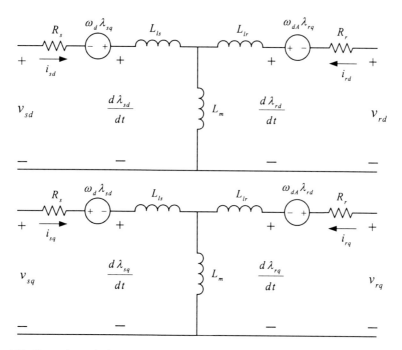

Fig. 4.19 Dynamic equivalent circuits of a DFIG with rotor injection

leading the d-axis by 90°. The dynamic equivalent circuits (for d-axis and q-axis) of a DFIG are redrawn in Fig. 4.19. For the chosen reference frame,

$$\begin{aligned} \omega_d &= \omega_s \\ \omega_{dA} &= \omega_s - \omega_m = \omega_{slip} \end{aligned} \tag{4.19}$$

For the design of feedback controllers for the rotor converter of a DFIG-based wind generator, it is convenient to use the reduced order model of a DFIG. The reduced order model neglects the transients in the stator flux linkages and assumes them to be constant. This is a valid assumption for the purpose of designing torque and reactive power controllers for the rotor side converter with relatively small variations in the grid voltage. With this assumption, the two stator voltage equations given in (1.9) are simplified to (4.20) and (4.21). The rotor voltage equations of (1.9), flux equations of (1.10), and torque equation of (1.11) are retained in the reduced order model.

$$v_{sd} = R_s i_{sd} - \omega_d \lambda_{sq} \tag{4.20}$$

$$v_{sq} = R_s i_{sq} + \omega_d \lambda_{sd} \tag{4.21}$$

In the dq reference frame-based analysis and control, there is a choice in the alignment of the d-or q-axis with an appropriate synchronously rotating space vector. Typically, either the stator flux orientation or the stator voltage orientation is used for the control of a DFIG. In stator flux-oriented control (FOC), the d-axis is aligned with the stator flux space vector $\vec{\lambda}_s$ resulting in $\lambda_{sq} = 0$ and $\lambda_{sd} = \left|\vec{\lambda}_s\right|$. In the stator voltage-oriented control (VOC), the d-axis is aligned with the stator voltage space vector $\vec{v}_s$, leading to (4.22) and (4.23). VOC has the advantage that it is relatively easier to obtain and align with the stator voltage space vector from the measured phase voltages. Also, the control of the grid side converter is typically done using stator voltage orientation, making it easier to use VOC for rotor side control as well. When the voltage drops across the stator resistances are neglected, the two orientations (FOC and VOC) become similar. In both orientations, it is possible to fully decouple the control of active power (or torque) and the reactive power. The analysis and controller design in the following sections are done based on stator voltage orientation.

$$v_{sd} = \sqrt{\frac{2}{3}}\, |\vec{v}_s|$$

$$v_{sq} = 0 \tag{4.22}$$

For balanced grid voltages, magnitude of $\vec{v}_s$ is $\sqrt{\frac{3}{2}}\,\hat{V}_{an}$, where $\hat{V}_{an}$ is the peak value of the phase to neutral voltage, and therefore, v_{sd} is also equal to the RMS value of the line–line voltage. Assuming that the voltage drop across the stator resistance and stator leakage inductance is small enough, and by neglecting the stator voltage transients, the stator flux space vector $\vec{\lambda}_s$ can be approximated to be always lagging the stator voltage space vector $\vec{v}_s$ by $90°$, i.e., aligned with the negative q-axis. This results in the approximate expressions of (4.23), which are also consistent with (4.20) and (4.21) assuming the resistive drops are negligible.

$$\lambda_{sd} \approx 0$$

$$\lambda_{sq} \approx -\left|\vec{\lambda}_s\right| \approx -\frac{v_{sd}}{\omega_s} \tag{4.23}$$

4.3.3 References for Rotor d- and q- Axes Currents

The electromagnetic torque, the active power injected into the grid, and the reactive power at the stator can all be controlled by controlling the injected rotor currents. With the assumptions used in (4.23), the VOC can be shown to provide decoupled control of active and reactive power, with i_{rd} controlling the active power and i_{rq} controlling the reactive power as derived below.

Using the current-flux linkage relationship given in (1.10), the stator dq currents i_{sd} and i_{sq} are obtained as (4.24) and (4.25).

$$i_{sd} = \frac{\lambda_{sd}}{L_s} - \frac{L_m}{L_s}i_{rd} \tag{4.24}$$

$$i_{sq} = \frac{\lambda_{sq}}{L_s} - \frac{L_m}{L_s}i_{rq} \tag{4.25}$$

The active power at the stator in the dq reference frame is given by (4.26).

$$P_s = v_{sd}i_{sd} + v_{sq}i_{sq} \tag{4.26}$$

For the VOC where $v_{sq} = 0$ and $\lambda_{sd} = 0$, (4.26) simplifies to (4.27).

$$P_s = v_{sd}i_{sd} = v_{sd}\left(\frac{\lambda_{sd}}{L_s} - \frac{L_m}{L_s}i_{rd}\right) = -\frac{L_m}{L_s}v_{sd}i_{rd} \tag{4.27}$$

As seen from (4.27), the stator active power is fully controlled by the rotor d-axis current, and is independent of the value of q-axis rotor current. The expression for the total electromagnetic torque is given in (1.11). For VOC, using (4.23–4.25), the torque expression simplifies to (4.28). Again, the control of torque is through the control of rotor d-axis current and is independent of the q-axis current.

$$\begin{aligned}
T_{em} &= \frac{p}{2}L_m\left(i_{sq}i_{rd} - i_{sd}i_{rq}\right) \\
&= \frac{p}{2}L_m\left(\left(\frac{\lambda_{sq}}{L_s} - \frac{L_m}{L_s}i_{rq}\right)i_{rd} - \left(\frac{\lambda_{sd}}{L_s} - \frac{L_m}{L_s}i_{rd}\right)i_{rq}\right) \\
&= -\frac{p}{2}\frac{1}{\omega_s}\frac{L_m}{L_s}v_{sd}i_{rd}
\end{aligned} \tag{4.28}$$

Next, we will consider the reactive power expressions. The reactive power at the stator in the dq reference frame is given by (4.29).

$$Q_s = v_{sq}i_{sd} - v_{sd}i_{sq} \tag{4.29}$$

For the VOC where $v_{sq} = 0$ and $\lambda_{sq} = -\dfrac{v_{sd}}{\omega_s}$, (4.29) simplifies to (4.30).

$$Q_s = -v_{sd}i_{sq} = -v_{sd}\left(\frac{\lambda_{sq}}{L_s} - \frac{L_m}{L_s}i_{rq}\right) = \frac{(v_{sd})^2}{\omega_s L_s} + \frac{L_m}{L_s}v_{sd}i_{rq} \tag{4.30}$$

Again, as seen from (4.30), the reactive power depends on v_{sd} and i_{rq} only, and is independent of i_{rd}.

Since the electromagnetic torque and the reactive power at the stator are controlled by controlling i_{rd} and i_{rq}, respectively, the current references i_{rd}^* and i_{rq}^* for the current control loops can be derived independently and are given by (4.31) and (4.32). The reference i_{rd}^* is derived from the torque command T_{em}^*, from the speed control loop along with the grid frequency support control loop (refer Fig. 4.17), and from (4.28). The reference i_{rq}^* is derived from the reactive power command Q_s^* from the stator terminal voltage regulation loop, or as a direct command for reactive power or power factor (refer Fig. 4.17), and from using (4.30).

$$i_{rd}^* = -\frac{2}{p}\omega_s \frac{L_s}{L_m}\frac{1}{v_{sd}}T_{em}^* \tag{4.31}$$

$$i_{rq}^* = \frac{L_s}{L_m}\frac{1}{v_{sd}}Q_s^* - \frac{v_{sd}}{\omega_s L_m} \tag{4.32}$$

4.3.4 Controller Design for Rotor Current Loops

The d-axis and q-axis rotor current loops are designed and implemented independently, and generate the references for the d-axis and q-axis rotor voltages. Based on these reference signals, the rotor side PWM converter generates the three-phase rotor voltages, typically using space vector PWM. In order to design the rotor current loops, the small-signal transfer function from the rotor voltage to the rotor current in each of d- and q- axes need to be derived.

From the equivalent circuit of Fig. 4.19, the d-axis voltage is given by (4.33).

$$v_{rd} = R_r i_{rd} + \frac{d}{dt}(L_m i_{sd} + L_r i_{rd}) - \omega_{slip}(L_m i_{sq} + L_r i_{rq}) \tag{4.33}$$

The assumption that the stator flux transients are negligible in the reduced order model used in previous sections implies (4.34).

$$\frac{d}{dt}\lambda_{sd} = \frac{d}{dt}(L_s i_{sd} + L_m i_{rd}) = 0$$

$$\tag{4.34}$$

$$\therefore \frac{d}{dt}i_{sd} = -\frac{L_m}{L_s}\frac{d}{dt}i_{rd}$$

Substituting (4.34) in (4.33),

$$v_{rd} = R_r i_{rd} + \sigma L_r \frac{d}{dt}i_{rd} - \omega_{slip}(L_m i_{sq} + L_r i_{rq}) \tag{4.35}$$

where, $\sigma = \left(1 - \frac{L_m^2}{L_s L_r}\right)$

Substituting for i_{sq} from (4.25) into (4.35) results in the required expression in (4.36), which is useful for deriving the transfer function from the d-axis voltage to current.

$$v_{rd} = R_r i_{rd} + \sigma L_r \frac{d}{dt}i_{rd} - \omega_{slip}\sigma L_r i_{rq} - \omega_{slip}\frac{L_m}{L_s}\lambda_{sq} \tag{4.36}$$

While obtaining the small-signal transfer function from the d-axis voltage to current, the last term in (4.36) is negligible since λ_{sq} is assumed constant and independent of changes in d-axis rotor current. Furthermore, the term $\omega_{slip}\sigma L_r i_{rq}$ in

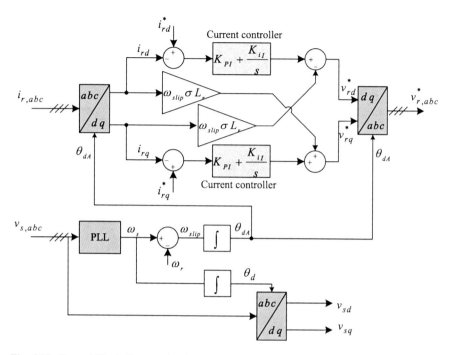

Fig. 4.20 Control block diagram for the rotor current control

(4.36) is fed forward as shown in the control block diagram of Fig. 4.20, which leads to the transfer function given in (4.37).

$$\frac{I_{rd}(s)}{V_{rd}(s)} = \frac{1}{R_r + s\sigma L_r} \tag{4.37}$$

Similar to the derivation of (4.36), the q-axis rotor voltage can be expressed as

$$v_{rq} = R_r i_{rq} + \sigma L_r \frac{d}{dt} i_{rq} + \omega_{slip}\sigma L_r i_{rd} + \omega_{slip}\frac{L_m}{L_s}\lambda_{sd} \tag{4.38}$$

Again, assuming zero perturbation in λ_{sd} and feeding forward the term $\omega_{slip}\sigma L_r i_{rd}$ the transfer function for the q-axis control loop can be obtained as (4.39).

$$\frac{I_{rq}(s)}{V_{rq}(s)} = \frac{1}{R_r + s\sigma L_r} \tag{4.39}$$

A proportional-integral (PI) controller for each of the two control loops can be easily designed for the simple plant transfer functions given in (4.37) and (4.39) for a desired bandwidth and phase margin. Since the plant transfer functions for the two loops are similar, their controllers will also be similar.

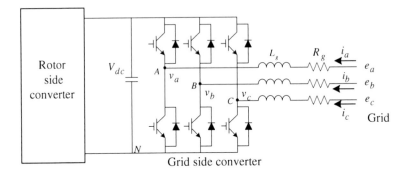

Fig. 4.21 Schematic of the grid side converter

4.3.5 *Control of the Grid Side Converter*

The main role of the grid side converter is to provide a path for the active power exchange in the positive or negative direction between the rotor side converter and the grid. It does so by regulating the DC link voltage. In the subsynchronous mode when the RSC absorbs power from the DC link, the GSC draws power from grid and sources to the DC link, and in the supersynchronous mode, the GSC reverses power direction and injects the power from the RSC into the grid. In steady-state, the DC current drawn by the GSC equals the DC current injected by the RSC. The reactive power of GSC can be controlled independently of the active power exchanged. Due to the DC link between the two converters, the reactive power processed by the GSC is also independent of the reactive power processed by the RSC. This capability is used to provide additional reactive power support to the grid within the ratings of the converter similar to the operation of a STATCOM, even when the turbine is not rotating.

4.3.5.1 Analysis of GSC in the Synchronous Frame

The schematic of the GSC interfacing the DC link and the three-phase grid and comprising an inductive filter L_g with a winding resistance of R_g is shown in Fig. 4.21. Similar to RSC, the control of GSC is also conveniently done in the synchronously rotating frame aligned with the grid voltage space vector. The active power and reactive power control are decoupled in this voltage-oriented control (VOC). In order to derive the synchronous dq frame equations, we define the grid voltage vector, $\underline{e}$, converter phase voltage (average) vector, $\underline{v}$, and phase current vector, $\underline{i}$ in (4.40).

$$\underline{e} = \begin{bmatrix} e_{an} \\ e_{bn} \\ e_{cn} \end{bmatrix} \quad \underline{v} = \begin{bmatrix} v_{an} \\ v_{bn} \\ v_{cn} \end{bmatrix} \quad \underline{i} = \begin{bmatrix} i_a \\ i_b \\ i_c \end{bmatrix} \quad (4.40)$$

Note that cycle-by-cycle average value of each of the pole output voltages contains a DC component of $V_{dc}/2$ and a controlled AC component at the grid frequency. The voltage vector $\underline{v}$ represents this AC component in each of the pole voltages. Also, note that the neutral voltage with respect to the DC negative (N in Fig. 4.19) under balanced conditions is $V_{dc}/2$, i.e., $v_{nN} = V_{dc}/2$.

The transformation matrix, $\underline{T}$ for converting abc quantities to the dq frame is given by,

$$
\begin{bmatrix} x_d \\ x_q \end{bmatrix} = \sqrt{\frac{2}{3}} \underbrace{\begin{bmatrix} \cos(\omega_s t) & \cos\left(\omega_s t - \frac{2\pi}{3}\right) & \cos\left(\omega_s t - \frac{4\pi}{3}\right) \\ -\sin(\omega_s t) & -\sin\left(\omega_s t - \frac{2\pi}{3}\right) & -\sin\left(\omega_s t - \frac{4\pi}{3}\right) \end{bmatrix}}_{T} \begin{bmatrix} x_a \\ x_b \\ x_c \end{bmatrix} \tag{4.41}
$$

The current and the two voltage vectors in the dq frame are denoted by (4.42).

$$
\underline{e}_{dq} = \begin{bmatrix} e_d \\ e_q \end{bmatrix} \quad \underline{v}_{dq} = \begin{bmatrix} v_d \\ v_q \end{bmatrix} \quad \underline{i}_{dq} = \begin{bmatrix} i_d \\ i_q \end{bmatrix} \tag{4.42}
$$

It is straightforward to establish the decoupled control of active power P_g through control of i_d, and reactive power Q_g through i_q in the grid voltage-oriented control, where e_q is zero and e_d equals $\sqrt{3/2}$ times the amplitude of the grid phase voltage.

$$
P_g = e_d i_d + e_q i_q = e_d i_d \tag{4.43}
$$

$$
Q_g = e_q i_d - e_d i_q = -e_d i_q \tag{4.44}
$$

The reference for the active power control loop comes from the DC link voltage controller, and the reference for the reactive power control loop comes from a system-level reactive power demand or power factor command. The references for the d- and q-axes GSC currents can then be obtained from (4.43) and (4.44). The transfer function from the d- and q-axes voltages to the respective currents needed for the design of current loops is derived below. The current loops generate the voltage reference vectors for the pulse-width modulator of GSC. From Fig. 4.21,

$$
\underline{e} = L_g \frac{d}{dt} \underline{i} + R_g \underline{i} + \underline{v} \tag{4.45}
$$

Multiplying both sides of (4.45) by the transformation matrix $\underline{T}$,

$$
\underline{T}\,\underline{e} = L_g \underline{T} \frac{d}{dt} \underline{i} + R_g \underline{T}\underline{i} + \underline{T}\underline{v} \tag{4.46}
$$

$$
\underline{T}\frac{d}{dt}\underline{i} = \frac{d}{dt}\left[\underline{T}\underline{i}_{dq}\right] - \left[\frac{d}{dt}\underline{T}\right]\underline{i} = \frac{d}{dt}\underline{i} - \omega_s \begin{bmatrix} i_q \\ -i_d \end{bmatrix} \tag{4.47}
$$

From (4.46) and (4.47),

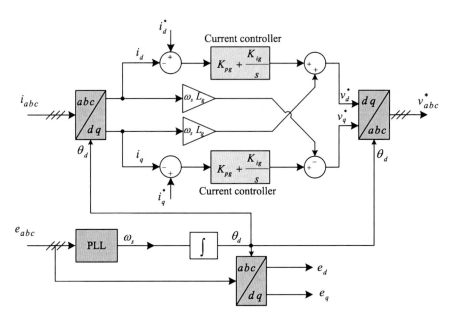

Fig. 4.22 Block diagram of grid side converter control

$$\underline{e}_{dq} = L_g \frac{d}{dt} \underline{i}_{dq} - \omega_s L_g \begin{bmatrix} i_q \\ -i_d \end{bmatrix} + R_g \underline{i}_{dq} + \underline{v}_{dq} \qquad (4.48)$$

Rewriting (4.48) in terms of the individual d and q components,

$$L_g \frac{d}{dt} i_d = e_d - R_g i_d - v_d + \omega_s L_g i_q \qquad (4.49)$$

$$L_g \frac{d}{dt} i_q = e_q - R_g i_q - v_q - \omega_s L_g i_d = -R_g i_q - v_q - \omega_s L_g i_d \qquad (4.50)$$

Similar to the technique adapted in the control of RSC currents, the cross-coupling terms $\omega_s L_g i_q$ and $\omega_s L_g i_d$ can be fed forward as shown in the control block diagram of Fig. 4.22. This leads to a simple first-order expression for the transfer functions from converter voltages to line currents given in (4.51) and (4.52) neglecting the transients in the grid voltage.

$$\frac{i_d(s)}{v_d(s)} = \frac{-1}{R_g + sL_g} \qquad (4.51)$$

$$\frac{i_q(s)}{v_q(s)} = \frac{-1}{R_g + sL_g} \qquad (4.52)$$

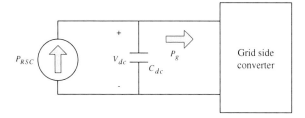

Fig. 4.23 Power balance at the DC link

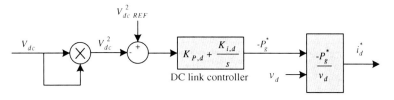

Fig. 4.24 Block diagram of DC link voltage control

4.3.5.2 DC Link Voltage Control

A simple scheme to regulate the DC link voltage in DFIG applications that results in a linear model is to control the square of the DC link voltage. For the purpose of designing the DC link voltage controller, the power from RSC into the DC link can be considered a constant, P_{RSC}, as shown in Fig. 4.23. P_g in Fig. 4.23 corresponds to the power flowing in to the GSC. Based on the energy stored in the DC link capacitor, (4.53) can be obtained.

$$\frac{1}{2} C_{dc} \frac{d}{dt} \left(V_{dc}^2 \right) = P_{RSC} - P_g \tag{4.53}$$

Treating V_{dc}^2 as the control variable and considering the perturbation in P_{RSC} to be zero, the small-signal model from (4.53) is derived as (4.54), which can be compensated by a proportional or PI controller.

$$\frac{V_{dc}^2(s)}{-P_g(s)} = \frac{2}{s\,C_{dc}} \tag{4.54}$$

The control structure for the DC link voltage regulation where the square of the measured DC link voltage is compared with a reference to generate active power reference, and in turn the reference for $i_d^* = -P_g^*/e_d$ is shown in Fig. 4.24.

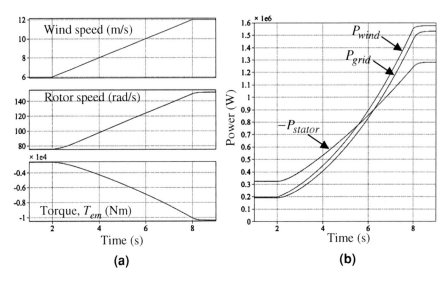

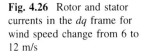

Fig. 4.25 Simulation plots corresponding to wind speed change from 6 to 12 m/s. **a** wind speed, rotor speed, and torque, **b** wind, grid, and stator power

Fig. 4.26 Rotor and stator currents in the dq frame for wind speed change from 6 to 12 m/s

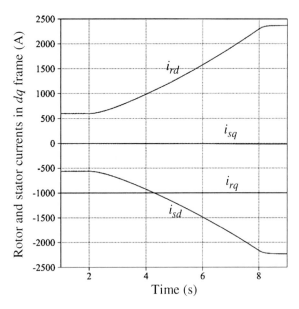

4.3.6 Simulation Results

Results from numerical simulation of a DFIG system with the parameters given in Table 4.1 and with controllers designed as per the methods outlined in previous sections are presented here. The simulation results validate the controller design

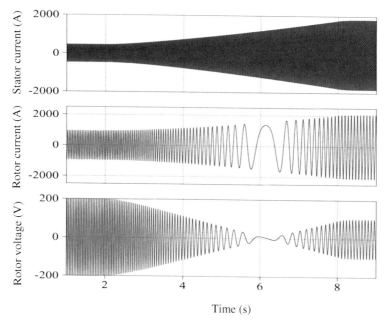

Fig. 4.27 Stator current, rotor current, and rotor voltage in phase-a for a wind speed change from 6 to 12 m/s

methods, and more importantly give better insight into the dynamic operation of the DFIG wind systems. The initial set of simulations corresponds to a linear change in the applied wind velocity from 6 to 12 m/s over a period of 6 s (from $t = 2$ to $t = 8$ s). The speed controller optimizes the rotor speed such that maximum possible power is captured at every instant. The reactive power command is set at zero. Figure 4.25a shows the corresponding plots of wind velocity, rotor speed (along with the rotor speed reference), and the electromagnetic torque. As seen, the rotor velocity also changes in step with the wind velocity to keep the tip speed ratio constant, and the power coefficient at its maximum value. Figure 4.25b shows the plots of wind power, power injected into the grid, and the stator power. As seen, wind power matches with the wind model, and the grid power tracks the wind power closely. The difference between the grid power and the wind power corresponds to the losses in the system as well as to the increase in the stored energy of the turbine with increasing speed. The stator power is less than the total power for negative slip and higher than the total power for positive slip; the difference corresponds to the rotor converter power in each case.

Figure 4.26 shows the stator currents and the rotor currents in the dq frame. The rotor currents i_{rd} and i_{rq} are directly controlled by the torque and reactive power control loops, respectively, and the stator currents are in response to these rotor current injections. The value of i_{rq} corresponds to $-v_{sd}/(\omega_s L_s)$ for zero reactive power command, and i_{sq} being zero, confirms unity power factor operation. Also, note that both i_{rq} and i_{sq} are independent of the wind speed and remain constant.

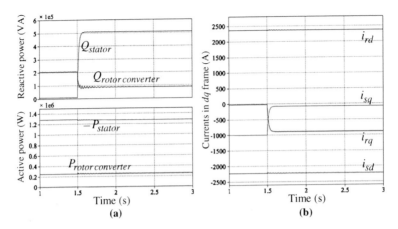

Fig. 4.28 Response to a step change in reactive power command by 0.5 MVAR **a** stator and rotor converter reactive power, and stator and rotor converter active power, **b** stator and rotor currents in the dq frame

Figure 4.27 shows the phase-a stator current, rotor current, and the rotor voltage in the abc reference frame. The stator current magnitude increases with the wind speed corresponding to higher power output, and is always at the grid frequency of 60 Hz. The rotor current magnitude also increases with the wind speed due to higher power processed, but at the slip frequency of $s*60$ Hz. The rotor voltage is roughly proportional to $s V_s$; hence the magnitude changes with $|s|$ starting at a value corresponding to 0.3 at the wind speed of 6 m/s, reducing to zero, and increasing back to 0.3 at 12 m/s. The frequency also follows s and is close to zero at the synchronous operation around t = 6 s. It can also be verified that all the plots match well with those obtained from the phasor analysis in Sect. 4.2.

Simulations corresponding to changes in the reactive power command are presented next. A step change of 0.5 MVAR in the reactive power command is applied with the wind speed kept constant at 12 m/s. Figure 4.28a shows the corresponding plots of the reactive power in the stator and rotor converter. Figure 4.28b shows the corresponding stator and rotor currents in the dq frame. As seen, only the q-axis currents change in response the Q command. The small disturbances in the d-axis current are due to the imperfect decoupling of grid voltage-oriented control, but as seen from the plots, these are negligible.

References

1. Peña R, Clare JC, Asher GM (1996) Doubly fed induction generator using back-to-back PWMPWM converters and its application to variable speed wind-energy generation. Proc Inst Elect Eng Elect Power Appl 143(3):231–241

2. Muller S, Deicke M, De Doncker RW (2005) Doubly fed induction generator systems for wind turbines. IEEE Ind Appl Mag 8(3):26–33
3. Miller NW, Price WW, Sanchez-Gasca JJ (2003) Dynamic modeling of GE 1.5 and 3.6 wind turbine-generators. GE-power systems energy consulting, General Electric International, Inc., Schenectady, NY, USA, 27 Oct 2003
4. Ellis G, Lorenz RD (2000) Resonant load control methods for industrial servo drives. Proc IEEE Ind Appl Soc Conf 3:1438–1445
5. Qiao W (2009) Dynamic modeling and control of doubly fed induction generators driven by wind turbines. Proceedings of the IEEE power systems conference and exposition (PSCE), 2009

Chapter 5
Dynamic Models for Wind Generators

5.1 Introduction

In the previous chapters some details regarding the types of wind generators, converter configurations used in wind turbine generators, and the nature of controllers used in the wind generators were provided. In this chapter, a description of the mathematical models used for the time domain simulation of the electro-mechanical phenomena associated with wind turbine generators will be provided. A discussion regarding the suitability of these models for time domain simulation will also be presented.

Several informative papers and reports dealing with modeling of wind generation for the purpose of power system dynamic performance analysis and control exist, as well as papers and presentations from an all-day panel session on wind generator modeling and controls for power system dynamics [1–9]. This panel session consisted of nine papers that covered a range of topics associated with models for wind turbine generators and their controls. A CIGRÉ working group report [10] also provides an excellent documentation of a range of materials associated with modeling and control of wind generators. Two other committee papers dealing with generic models for wind generators [11] and model validation [12] provide valuable information regarding generic models for simulation and the performance of these models under different situations.

Specific aspects of modeling requirements for power flow analysis and for transient stability analysis will be addressed here. In addition, some aspects of wind farm modeling will also be discussed.

V. Vittal and R. Ayyanar, *Grid Integration and Dynamic Impact of Wind Energy*,
Power Electronics and Power Systems, DOI: 10.1007/978-1-4419-9323-6_5,
© Springer Science+Business Media New York 2013

5.2 Modeling of Wind Turbine Generators for Power Flow Analysis

The power flow solution represents the starting point for the transient stability simulation by determining the pre-disturbance stable operating point. Accurate representation of the wind turbine generators for power flow analysis is an important step in determining the appropriate steady-state solution. The primary objective of the steady-state analysis conducted using power flows is to ascertain that the steady-state voltage, either pre-disturbance or post-disturbance, meets the necessary reliability standards. Typically, these limits are $\pm 10\ \%$ of nominal voltage. Given this range of allowable system voltage variation, it is to be noted that most modern wind turbines will operate continuously in this range without any low or high voltage tripping. Hence, for power flow analysis there is no necessity for representing low or high voltage tripping of wind turbines. The approach to represent each type of wind turbine generator in the power flow solution will now be discussed.

For Type 1 and Type 2 wind turbine generators the most accurate approach to modeling these machines in the power flow would be to incorporate their equivalent circuit in the power flow network model. Most positive sequence power flow software packages do not have this capability. Hence, a practical approach to model Type 1 and Type 2 wind turbine generators in the power flow is to represent the generator bus as a P-Q bus with a constant reactive power given by the amount of reactive power being absorbed at the level of real power generated by the machine for the study being conducted. In actuality, the reactive power consumption will change with the change in terminal voltage from the nominal value. This change is dependent on the characteristics of the machine. Since the steady-state equivalent circuit model of the machine is not represented in the positive sequence power flow, it is not possible to determine the reactive power consumption as a function of the active power and terminal voltage.

It is important that once the reactive power consumption of the machine has been determined, the shunt capacitors at the wind turbine compensating this reactive power consumption should be explicitly modeled and not lumped for the whole unit as a P-Q bus at a specified power factor. The shunt compensation calculated at the initial starting point of the power flow could provide adequate reactive support to compensate for the reactive power consumption in the machine and could be controlled using switching increments of shunt capacitors to regulate the effective power factor as the machine output changes. It should, however, be noted that this representation is not accurate since it only holds for the initial voltage magnitude at which the compensation was calculated.

As an example, consider the case of representing a machine operating at full load in the power flow. In this case, all the rated shunt compensation would be pressed into service at the rated voltage. If the system is now examined for changing conditions or for contingencies, the reactive compensation provided by the shunt compensation in service will vary with the square of the voltage

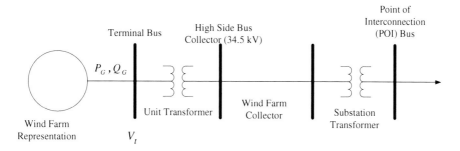

Fig. 5.1 Representation of wind farms for steady-state power flow analysis

magnitude, while that of the wind turbine generator could only be accurately determined if the steady-state equivalent circuit of the machine was included. Hence, a fixed power factor P-Q bus representation of the machine could provide optimistic results for analyzing steady-state voltage performance.

For Type 3 and Type 4 wind turbine generators, the units are typically modeled in power flows as a P-V bus with the rated VAr limits. Both these types of units have reactive power capability because of the converters being present. For the full converter Type 4 machines it is important to note that the machine should be treated as a constant current device rather than the conventional constant power device when the machine reaches it reactive power limit. In this case the VAr limit will change linearly with voltage magnitude once the unit reaches its reactive power limit. In such cases the accurate model consists of representing unit as a current source in the power flow once the reactive power limit is reached. Such features are typically not available in most power flow software packages.

In analyzing the impact of a wind farm on system steady-state performance, it is not essential to model each wind turbine in a wind farm. The wind farm output can be equivalenced. However, the underlying network structure that represents the voltage levels within the wind farm and the substation transformer at the point of interconnection to the bulk power system should be represented. Figure 5.1 shows a schematic of a typical wind farm representation for steady-state power flow analysis.

Another important aspect of steady-state analysis is short circuit analysis. In contrast to power flow analysis where the primary intent is to examine the impact of the wind farm as a whole on system performance, the intent of short circuit analysis is to examine the nature and magnitude of currents due to short circuits. In this situation the aggregate model for the wind farm normally used in power flow and time domain dynamic analysis would not provide the requisite detail if the intent is to examine the impact of the short circuit on the wind farm components. In this case a significant amount of detail regarding the wind farm feeder circuits and individual wind turbine units would need to be represented [13]. The type of wind turbine generator will also impact the contribution to the short circuit current.

For Type1 and Type 2 wind turbine generators the short circuit current characteristics are similar. For these types of wind turbine generators the short circuit representation in a typical fault analysis software package includes the induction generator as a voltage behind subtransient reactance. For unbalanced faults, it is generally assumed that both the negative and positive sequence impedances are equal to the subtransient reactance. It should also be noted that wind turbine generators are normally not grounded. Hence, there is no contribution from ground currents. Wind farms, however, have grounding transformers and power transformers that provide a path to ground and should be included in the detailed wind farm representation for short circuit studies.

Type 4 wind turbine generators that include a voltage-source converter produce a voltage behind the output inductance. The magnitude and phase of this voltage determines the real and reactive power output of the machine. This voltage, however, is highly controllable because of the pulse-width modulation used. In addition, the power electronic components used in the converters are susceptible to high current magnitudes, and the fast control inherently available can be utilized to limit the short circuit currents in Type 4 wind turbine generators. The ability to control or limit the fault currents in Type 4 wind turbine generators is dependent on the specific proprietary control provided by the manufacturer; the performance in limiting fault current can vary significantly from one design to another. Hence, the conventional short circuit modeling representation of a fixed voltage behind fixed impedance is not accurate for Type 4 machines. For unbalanced faults, the representation of these complex controls is needed to accurately capture the phenomenon. In such cases detailed transient simulation using tools such as PSCAD or EMTP must be performed and the power converters and controls need to be modeled detailing the manufacturer's design and control procedure.

The characteristics of Type 3 wind turbine generators are a combination of a Type 4 machine and an induction generator. As described in Chap.3, Sect. 3.2.3, a crowbar circuit is used to short the rotor and limit the extreme voltage and current duty placed on the converter because of the fault. The doubly fed machine responds like an induction generator when the crowbar is deployed, and when the crowbar is not deployed, the fault current contribution is controlled like a Type 4 machine.

5.3 Modeling of Wind Turbine Generators
for Transient Stability Analysis

For any given type of wind turbine, certain key elements that impact the dynamic performance of the wind turbine generator have to be included in the mathematical model representing the wind turbine generator in time domain simulation. This is required to accurately capture the dynamic behavior of the wind turbine generator and its impact on the system. The key elements that need to be represented in the model include:

• Wind turbine aerodynamic characteristics. This was described in detail in Chap.1

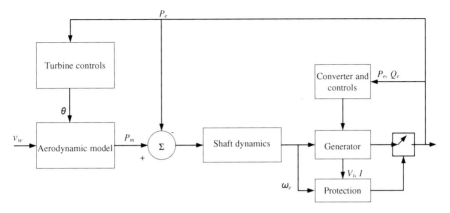

Fig. 5.2 Components of a wind turbine generator

- Wind turbine mechanical controls—pitch control and active stall control of the mechanical power delivered to the shaft. These controls were described in Chap. 4
- Mechanical shaft dynamics. This was detailed in Chap. 1. The model consists of a two-mass shaft. One mass represents the rotor/turbine blades and the second mass represents the generator.
- Electrical generator characteristics. This was also described in Chap. 1.
- Electrical controls—converter controls.
- Settings for protection.
- Transducers for measuring critical variables.

Figure 5.2 details the various components of the wind turbine generator model and the interaction between the various components.

A discussion of the modeling requirements for the various types of wind generators is presented in [10]. Although the parameters for the aerodynamics, turbine controls, and protection systems for the various types of wind turbine generators may differ from one manufacturer to another, the model structure will essentially be similar. In the case of stall controlled units, there is no turbine blade pitch control. For constant speed units, blade pitch will be used to regulate power instead of speed on variable speed units.

Since Type 3 wind turbine generators are the most widely used class of wind turbine generators, a detailed presentation of the modeling requirement for this type of wind turbine generator is presented in what follows. Three main controllers provide controls for frequency/active power, voltage/reactive power, and pitch angle/mechanical power. The outputs from these three controllers would provide inputs to the end block of the Type 3 machine completing the model representing the wind turbine generator.

In setting up a typical time domain transient stability study, the power flow would provide the operating condition for the Type 3 machine under consideration. The real power output, reactive power output, and the voltage at the terminal bus from the power flow are used to initialize the model, and the outputs from the model provide the current injection into the network at the generator bus.

The electrical power output (P_e) of the wind generator is then utilized in calculating the reference rotor speed (ω_{ref}) of the wind turbine. The reference rotor speed at rated wind speed is dependent on the design. Typically, it is about 1.2 pu at rated wind speed. However, for lower power output levels, the reference speed bears a nonlinear relationship with the electrical power output of the machine and is given by the general relationship

$$\omega_{ref} = f(P_e) \tag{5.1}$$

The nonlinear relationship shown above is manufacturer dependent. This relationship typically is only provided by the manufacturer.

5.3.1 Aerodynamic Model

The aerodynamic or wind power model for Type 3 wind turbine generators was developed in Sect. 1.3.3. This model describes the mechanical power extraction from the wind for a given pitch angle (θ) and a given tip speed ratio (λ). The aerodynamic or wind power model, which is turbine dependent computes the mechanical power extracted from the wind as shown in Eqs. (1.7) and (1.8), repeated here for convenience.

$$P_m = \frac{1}{2}\rho v_w^3 \pi r^2 C_p(\lambda) \tag{5.2}$$

$$\lambda = \frac{\omega_t r}{v_w} \tag{5.3}$$

The aerodynamic model of the wind turbine functions in conjunction with the active power and pitch angle controllers. This interaction is depicted in Fig. 5.3.

5.3.2 Mechanical Control and Shaft Dynamics

As illustrated in Fig. 5.3, the inputs to the aerodynamic model are the wind speed (v_w), pitch angle (θ), and turbine mechanical speed (ω_t). The mechanical power extracted is controlled by the pitch angle controller. For wind speeds beyond rated value, the blades of the wind turbine are pitched to limit the mechanical power delivered to the shaft to the equipment rating (1.0 pu). When the wind speed is less than the rated value, the blades are set at minimum pitch to maximize the mechanical power.

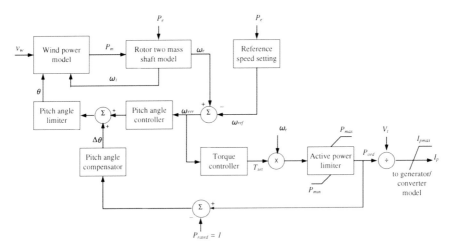

Fig. 5.3 Schematic diagram of active power and pitch angle controllers of Type 3 machines

The shaft dynamics is modeled as two masses as depicted in Sect. 1.3.4, taking into account separate masses for the turbine blades and the electrical generator. The model thus evaluates the generator rotor speed (ω_e) and the turbine mechanical speed (ω_t). The deviation of rotor speed from the reference speed (ω_{ref}) actuates the pitch angle controller and torque control modules simultaneously.

In addition to actuating the pitch angle controller in response to the speed error (ω_{err}), Type 3 machines have an additional block referred to as "pitch compensation," which provides a pitch angle error signal in response to the deviation of the output power from the rated value. When the electrical power output increases beyond the rated value, the pitch compensator acts to increase the pitch angle and returns the power back to the rated value.

The torque command (T_{set}) is used to compute the active power order (P_{ord}), which in turn provides the excitation current to the rotor side converter. In other words, the turbine control model sends a power order signal to the electrical control requesting that the converter deliver this power to the grid. The maximum active power order (P_{max}) from the controller is limited by the active power limiter block shown in Fig. 5.3. The current command (I_p) is computed by dividing P_{ord} from the wind turbine model by the generator terminal voltage (V_t). The current command is limited by the short-term current capability of the converter (I_{pmax}).

5.3.3 Electrical Generator Characteristics

The conventional wound rotor induction generator model was described in Sect. 1.3.5. In a Type 3 wind turbine generator, the fundamental frequency electrical dynamic performance is dominated by the converter. Aspects related to conventional generator performance such as internal angle, excitation voltage, or synchronism are not relevant. The rotor side converter drives the rotor current very fast and as a result the rotor flux dynamics are neglected. The electrical performance of the generator coupled with the inverter is akin to that of a current-regulated, voltage-source inverter [10]. Similar to other voltage-source inverters, the wind turbine generator with the converter can be represented as an internal voltage behind a transformer reactance delivering the desired active and reactive current to the device terminals.

This model does not contain any mechanical state variables describing rotor dynamics. The rotor dynamics are captured by the two-mass shaft dynamics model. All flux dynamics are neglected because the dynamics are dominated by the rapid response due to the electrical controls through the converter. This results in a controlled current source represented purely by algebraic equations. The controlled current source derives inputs in response to flux and active current commands from an electrical control model, and based on these commands derives the required inject current into the network. This model also represents time lags in converter action. Comparison of this simplified model with more detailed models of the generator and controls has demonstrated the efficacy of this model [10].

The models associated with different brands of wind turbines would typically have manufacturer-specific proprietary details. In order to facilitate analysis of dynamic phenomena associated with increased wind generator penetration, the IEEE Power and Energy Society's Power System Dynamic Performance Committee has formed a task force, which has developed generic models capturing the primary features that affect the dynamic performance of wind turbine generators and the associated controls [11, 12]. These generic models will be discussed in Sect. 5.3.5.

5.3.4 Electrical Control

The electrical control model provides a simplified mathematical representation of the converter control system. The model determines the active power and the reactive power to be delivered to the system based on inputs from the wind turbine model and the supervisory reactive power controller. The supervisory reactive power controller representation in the model provides reactive power control for the entire wind plant. This is typically implemented by monitoring a specified bus voltage and deriving an error using a reference voltage. A proportional-integral (PI) controller is used. The controller accounts for delays associated with cycle time, communication delays associated with each individual turbine, filtering in the controls, and voltage measurement lag. The supervisory reactive power controller provides a reactive power order, which in turn, is used by another controller to develop the reactive power control signal. This is then compared with the reactive power generated to determine the q-axis voltage command supplied to the generator/converter model.

The other loop within the electrical control model derives an input of the power order from the wind turbine model and the terminal voltage to derive the current

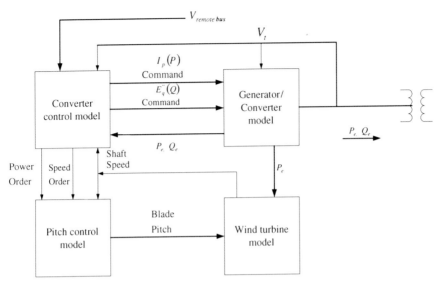

Fig. 5.4 Overall structure of Type 3 wind turbine generic model

command, which is also supplied to the generator/converter model. This control is well represented in the generic wind turbine models that have been developed, and will be presented in the following section.

5.3.5 Generic Model for Type 3 Wind Turbine Generators

As indicated in Sect. 4.3.3, a task force of the IEEE Power and Energy Society has led an initiative to develop generic models of wind turbine generators that do not contain proprietary information from the various manufacturers [11, 12]. The model developed for Type 3 machines is now presented. The models for the various elements of the Type 3 machines follow the descriptions provided in Sects. 5.3.1–5.3.4, but depict the details associated with the various controllers. The models for the various elements are shown in Figs. 5.4, 5.5, 5.6, 5.7, 5.8, 5.9.

Figure 5.4 depicts the overall structure of the Type 3 wind turbine generic model. It includes four elemental blocks and the interactions associated with each block.

The generic wind turbine model for Type 3 machines is shown in Fig. 5.5. It should be noted that this model obtains an input of the blade pitch angle from the pitch control model Fig. 5.6 and provides the turbine mechanical speed as an input to the pitch control model and the active power model Fig. 5.9. It includes a simplified aerodynamic model, which calculates the mechanical power developed by the wind turbine using the blade pitch angle and the wind speed.

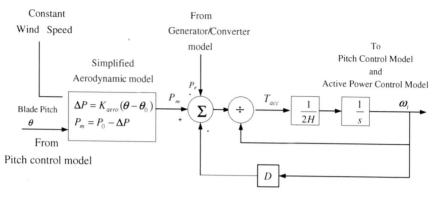

Fig. 5.5 Generic Type 3 wind turbine model

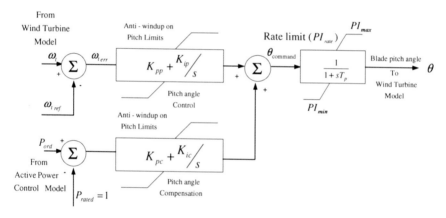

Fig. 5.6 Generic Type 3 pitch control model

The pitch control model is depicted in Fig. 5.6. The actuators associated with the blade position are rate limited, hence, a time constant is also included to represent the translation of the blade angle into mechanical output. Two PI controllers are represented in the blade pitch angle control. These controllers receive inputs of speed and power errors to synthesize the control signal.

The generic Type 3 generator/converter model is shown in Fig 5.7. In this model the flux dynamics are neglected to reflect the rapid response of the converter to the higher level commands arising from the electrical controls, which are depicted separately in Figs 5.8 and 5.9. The generic generator/converter model also includes low-voltage power logic (LVPL), which is used to limit the real current command during and immediately following sustained faults. The electrical controls associated with the converter are shown in Figs. 5.8 and 5.9. These controls include the reactive power control module shown in Fig 5.8 and the active power control module shown in Fig 5.9. The modules depict the controls

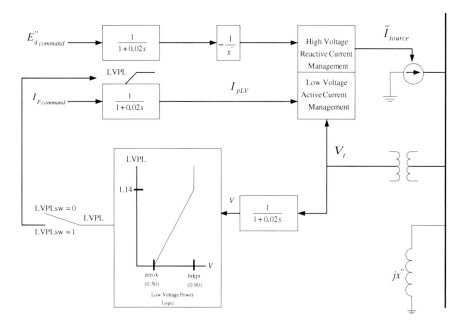

Fig. 5.7 Generic Type 3 generator/converter model

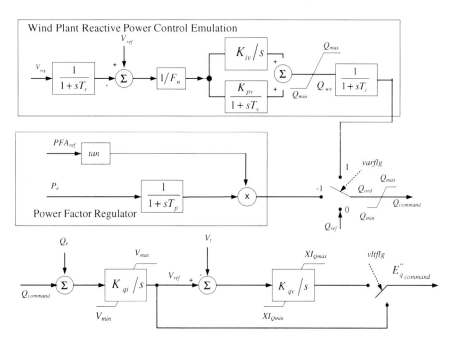

Fig. 5.8 Generic Type 3 reactive power control model

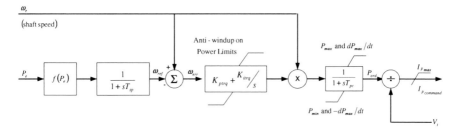

Fig. 5.9 Generic Type 3 active power control model

associated with the converter that determine the reactive power and the active power that will be delivered to the system. These quantities are determined by the signals $E''_{q\,command}$ and $I_{p\,command}$ respectively. In the control associated with the synthesis of the $E''_{q\,command}$ signal, the reactive power order Q_{ord} is computed by two separate control models representing the wind plant reactive power control emulator or the power factor regulator. The wind plant reactive power control emulator is a simplified representation of the supervisory VAr control element of the entire wind farm management system. The reactive power order Q_{ord} can also be held constant at the value of Q_{ref} as shown in Fig. 5.8.

The active power order $I_{p\,command}$ is derived utilizing the wind turbine generator electrical output power and shaft speed as shown in Fig. 5.9. In obtaining the active power order, the speed reference ω_{ref} is derived from the turbine speed setpoint versus power output curve (5.1).

5.4 Wind Farm Representation

A wind farm will typically include several wind turbine generators. For conducting transient stability studies in which the primary purpose is to examine the impact of the wind farm on system dynamic performance of the system, it is not necessary to represent each wind turbine in the wind farm separately. Typically, the wind farm is represented by a single suitably-sized equivalent interconnected to the grid at the appropriate point of interconnection. The control models discussed in the earlier sections of this chapter are also included with appropriate parameters. An example is presented below to compare the performance of the equivalent model with the detailed representation of each wind turbine in a time domain study.

A sample wind farm shown in Fig. 5.10 is considered. All the wind turbine generators are assumed to be the GE 1.5 MW DFIG machines [14]. The wind farm shown in Fig. 5.10 consists of seven wind turbine generators. The sample data for the various feeders connecting the wind turbine generators, the pad mount transformers at each wind turbine, and the interconnection to the grid are provided in Table 5.1. All impedance data provided are on a 100 MVA base.

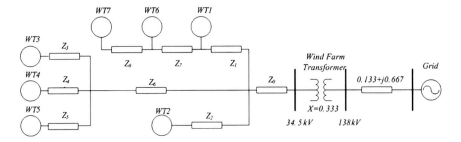

Fig. 5.10 Sample wind farm considered

Table 5.1 Wind farm parameters

Parameter	Value (p.u., based on 100 MVA)
Z_0	$0.0022 + j\,0.0119$
Z_1	$0.0040 + j\,0.0031$
Z_2	$0.0040 + j\,0.0031$
Z_3	$0.0073 + j\,0.0207$
Z_4	$0.0069 + j\,0.0187$
Z_5	$0.0097 + j\,0.0336$
Z_6	$0.0029 + j\,0.0156$
Z_7	$0.0040 + j\,0.0031$
Z_8	$0.0040 + j\,0.0031$
Pad-mount transformer Z_T	$0 + j\,3.2571$

In order to test the efficacy of a single machine equivalent representing the wind farm, a three-phase short circuit is placed at the point of the interconnection and cleared in four cycles. The simulation is conducted with all seven wind turbines represented as they exist in Fig. 5.10 and with the entire wind farm represented by a single equivalent. For the case where the seven wind turbines are represented in detail, the active and reactive power delivered to the grid and the voltage at the grid interconnection point are calculated at each time instant following the disturbance and compared with the corresponding quantities when the wind farm is represented by a single equivalent wind turbine generator. These plots are shown in Figs. 5.11, 5.12, 5.13. It is observed that with respect to the system at the interconnection point, the response of the equivalent machine accurately describes the aggregate behavior of the seven individual wind turbine generators. The point of interconnection behavior is critical in examining the impact of the wind farm on the rest of the system. These results demonstrate that the equivalent machine representation will accurately depict the impact of the wind farm with regard to system dynamic performance.

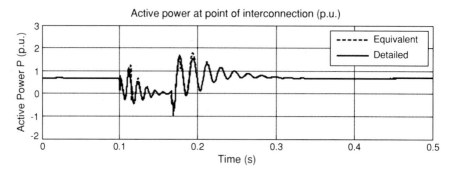

Fig. 5.11 Comparison of active power at point of interconnection

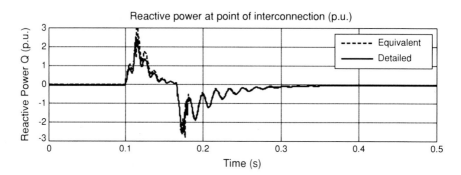

Fig. 5.12 Comparison of reactive power at point of interconnection

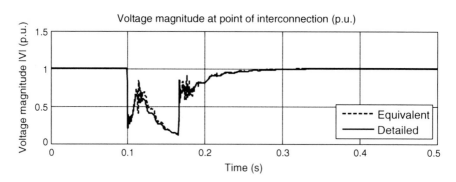

Fig. 5.13 Comparison of voltage magnitude at point of interconnection

References

1. Koessler RJ, Pillutla S, Trinh LH, Dickmander DL (2003) Integration of large wind farms into utility grids (Part 1—Modeling of DFIG). PES2003-000866, panel session on wind generator modeling and controls for power system dynamics, IEEE PES GM2003, Toronto, July 2003
2. Pourbeik P, Koessler RJ, Dickmander DL, Wong W (2003) integration of large wind farms into utility grids (Part 2—Performance Issues). PES2003-000926, panel session on wind generator modeling and controls for power system dynamics, IEEE PES GM2003, Toronto, July 2003
3. Kazachkov YA, Feltes JW, Zavadil R (2003) Modeling wind farms for power system stability studies. PES2003-000946, panel session on wind generator modeling and controls for power system dynamics, IEEE PES GM2003, Toronto, July 2003
4. Usaola J, Ledesma P, Rodriguez JM, Fernandez JL, Beato D, Iturbe R, Wilhelmi JR (2003) Transient stability studies in grid with great wind power penetration modeling issues and operation requirements. PES2003-000794, panel session on wind generator modeling and controls for power system dynamics, IEEE PES GM2003, Toronto, July 2003
5. Thiringer T, Petersson A, Petru T (2003) Grid disturbance response of wind turbines equipped with induction generator and doubly-fed induction generator. PES2003-000803, panel session on wind generator modeling and controls for power system dynamics, IEEE PES GM2003, Toronto, July 2003
6. Kabouris J, Vournas CD (2003) Designing controls to increase wind power penetration in weakly connected areas of the hellenic interconnected system. PES2003-000528, panel session on wind generator modeling and controls for power system dynamics, IEEE PES GM2003, Toronto, July 2003
7. Corsi S, Pozzi M (2003) Control systems of wind turbine generators an Italian experience. PES2003-000949, panel session on wind generator modeling and controls for power system dynamics, IEEE PES GM2003, Toronto, July 2003
8. Miller NW, Sanchez-Gasca JJ, Price WW, Delmerico RW (2003) Dynamic modeling of GE 1.5 and 3.6 MW Wind Turbine generators for stability simulations. PES2003-000590, panel session on wind generator modeling and controls for power system dynamics, IEEE PES GM2003, Toronto, July 2003
9. Palsson MP, Toftevaag T, Uhlen K, Tande JOG (2003) Control concepts to enable increased wind power penetration. PES2003-000534, panel session on wind generator modeling and controls for power system dynamics, IEEE PES GM2003, Toronto, July 2003
10. Cigré Report 328 (2007) Modeling and dynamic behavior of wind generation as it relates to power system control and dynamic performance. Working Group C4.601, Aug 2007
11. Adhoc TF on WTG Modeling, of the IEEE Working Group on Dynamic Performance of Wind Power Generation, "Description and Technical Specifications for Generic WTG Models—A Status Report. In: Proceedings of the power system conference and exposition (PSCE), 2011
12. Adhoc TF on WTG Modeling, of the IEEE Working Group on Dynamic Performance of Wind Power Generation, "Model Validation for Wind Turbine Generator Models. IEEE trans power sys, 26(3):1769–1782, Aug 2011
13. Samaan N, Zavadil R, Smith J, Conto J (2008) Modeling of wind power plants for short circuit analysis in the transmission network. In: Proceedings of the IEEE PES 2008 transmission and distribution conference and exposition
14. Miller NW, Price WW, Sanchez-Gasca JJ (2005) Modeling of GE wind turbine-generators for grid studies. General Electric International, Inc., Schenectady, Tech. Rep. version 3.4b Mar 2005

Chapter 6
Impact of Increased Penetration of DFIG Wind Generators on System Dynamic Performance

6.1 Introduction

In this chapter the impact of increased penetration of DFIG wind generators on system dynamic performance is examined. These impacts will be examined with respect to electromechanical phenomena associated with rotor angle stability, small-signal stability, voltage stability, and frequency stability. Among the several wind generation technologies, variable speed wind turbines utilizing doubly fed induction generators (DFIGs) are gaining prominence in the power industry. The performance of these machines is largely determined by the converter and the associated controls as described in Chap. 4. Since DFIGs are asynchronous machines, they have primarily four mechanisms by which they can affect the damping of electromechanical modes (since they themselves do not participate in the modes):

1. Displacing synchronous machines thereby affecting the modes.
2. Impacting major path flows thereby affecting the synchronizing forces.
3. Displacing synchronous machines that have power system stabilizers.
4. DFIG controls interacting with the damping torque on nearby large synchronous generators.

Two recent reports detail the impact of increased penetration of wind generators on two different portions of the North American interconnection [1, 2]. In these reports, significant aspects of the impact of increased wind penetration with regard to system dynamic performance are discussed along with steps that need to be taken to improve system reliability and operational efficiencies. The material in this chapter is motivated by the results presented in these reports and builds on the topics addressed in these reports to present specific details related to the impact of increased wind penetration on rotor angle stability, voltage stability, and frequency stability. These are three important aspects of power system dynamic performance as described in [3]. Each of these aspects will be considered separately and discussed.

V. Vittal and R. Ayyanar, *Grid Integration and Dynamic Impact of Wind Energy*,
Power Electronics and Power Systems, DOI: 10.1007/978-1-4419-9323-6_6,
© Springer Science+Business Media New York 2013

6.2 Impact on Rotor Angle Stability

In examining the impact of increased DFIG penetration on rotor angle stability two distinct aspects are discussed:

- Small-signal rotor angle stability
- Transient rotor angle stability

Each of these aspects differs with regard to the nature of the phenomenon being examined and the type of tools used to analyze them.

6.2.1 Impact on Small-Signal Rotor Angle Stability

In analyzing the impact of DFIG penetration on small-signal rotor angle stability a key assumption made is that the changes considered are small enough to warrant linearization of the system equations. Once this assumption is invoked, then all available tools to perform small-signal stability analysis using appropriate representation of the DFIG wind generators can be utilized to analyze the problem.

To motivate the impact of increased DFIG wind turbine generator penetration on small-signal rotor angle stability, consider the small power system example depicted in Fig. 6.1. This figure depicts a 9-bus system with three synchronous generators and a wind farm with an equivalent DFIG wind turbine generator connected to Bus # 9. The DFIG wind turbine is an asynchronous machine and injects real power and reactive power P_G and Q_G, respectively at Bus # 9. In response to this injection, the rotor angle positions of the three synchronous generators will adjust to achieve a new steady-state operating condition. Depending on the network impedances, loads, and the power flows on the existing transmission lines some of the rotor angle positions could advance and some could retard from their existing positions before the injection. In addition, this injection of power into the network is not accompanied by the addition of any inertia to the system since the inertia of the DFIG is completely masked due to the converters on the rotor side and grid side.

Given this setting, if a calculation of the synchronizing power capability of the multi-machine system following the steps outlined in [4, Chap. 3] is performed, then it would be observed that based on the change in rotor angle differences among the three synchronous generators in the system, the synchronizing capability of some of the machines would reduce if the relative rotor angle differences increase. Similarly, the synchronizing capability of some of the machines could increase if the relative rotor angle differences decrease. This would also affect the damping of the modes associated with these generators and the frequency of oscillation. Hence, the location of the injection from the wind farm and the magnitude of the power injection would have an impact on the system's oscillatory behavior.

Alternatively, the model developed for Type 3 wind turbine generators can be adopted and included in any conventional small-signal stability analysis software

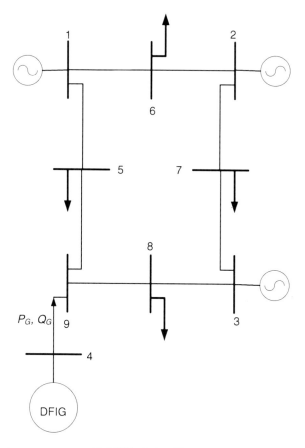

Fig. 6.1 Sample power system with DFIG wind turbine

package and the impact of the DFIG wind turbines on system oscillations can be systematically analyzed. Several recent publications follow this approach [5–10]. Using this approach for a specific system and considering actual locations and ratings of the wind farm(s), different operating conditions could be considered and the impact of the wind farm(s) on the small-signal stability behavior in terms of oscillation frequency and damping ratio could be evaluated. The formulation of the small-signal stability problem will be discussed in detail in the following section.

6.2.2 Formulation of the Small-Signal Stability Problem

The following general equations can be used to describe the dynamics of the power system,

$$\dot{x} = h(x, u, t) \tag{6.1}$$

$$y = g(x, u) \tag{6.2}$$

where

x = vector of state variables

h = system of nonlinear differential equations

u = vector of system inputs

y = vector of algebraic or network variables

For conducting small-signal stability analysis, the nonlinear equations of the dynamic power system are first linearized around a specific operating point. The linearized equations can be written in the form

$$\Delta \dot{x} = A \Delta x + B \Delta u \tag{6.3}$$

$$\Delta y = C \Delta x + D \Delta y \tag{6.4}$$

where, A, B, C, and D are known as the state or plant matrix, input matrix, output matrix, and feed forward matrix, respectively.

The state equations in the frequency domain can be obtained by applying the Laplace transformation to (6.3) as follows:

$$s \Delta X(s) - \Delta x(0) = A \Delta X(s) + BU(s) \tag{6.5}$$

$$(sI - A) \Delta X(s) = \Delta x(0) + BU(s) \tag{6.6}$$

Equation (6.6) can be transformed to make the origin the initial condition by a simple shift of variables; the zero input response of the system will be given by

$$(sI - A) \Delta X(s) = 0 \tag{6.7}$$

The values of s that satisfy (6.7) are known as the eigenvalues of matrix A. The eigenvalues of matrix A, thus, exhibit important information regarding system response to small perturbations and thus characterize the stability of the system. The time-dependent characteristic of a mode corresponding to an eigenvalue λ is given by $e^{\lambda t}$. A real positive eigenvalue determines an exponentially increasing behavior while a negative real eigenvalue represents a decaying mode. A complex eigenvalue with positive real part results in an increasing oscillatory behavior and one with a negative real part results in damped oscillation. The real component of the eigenvalue gives the damping, and the imaginary component gives the frequency of oscillation. The frequency of oscillation (f) and damping ratio (ξ) of a complex eigenvalue ($\lambda = \sigma + j\omega$) can be represented as

$$f = \frac{\omega}{2\pi} \tag{6.8}$$

$$\xi = \frac{-\sigma}{\sqrt{\omega^2 + \sigma^2}}. \tag{6.9}$$

The damping ratio determines the rate of decay of the amplitude of the oscillation. An eigenvalue of the state matrix A and the associated right eigenvector (v_i) and left eigenvector (w_i) are defined as

$$Av_i = v_i\lambda_i \tag{6.10}$$

$$w_iA = \lambda_iw_i. \tag{6.11}$$

The component of the right eigenvector gives the mode shape, that is, the relative activity of the state variables when a particular mode is excited. For example, the degree of activity of the state variable x_j in the ith mode is given by v_{ji} of right eigenvector v_i. The jth element of the left eigenvector w_i weighs the contribution of this activity to the ith mode. The participation factor of the jth state variable (x_j) in the ith mode is defined as the product of the jth component of the right and left eigenvectors corresponding to the ith mode,

$$p_{ji} = v_{ji}w_{ij}. \tag{6.12}$$

In large power systems, the small-signal stability problem can be either local or global in nature. Power system oscillation frequencies are generally in the range between 0.1 and 2 Hz depending on the number of generators involved. Local oscillations lie in the upper part of the range and consist of the oscillation of a single generator or a group of generators against the rest of the system. Stability (damping) of these oscillations depends on the strength of the transmission system as seen by the power plant, generator excitation control systems, and plant output. In contrast, inter-area oscillations are in the lower part of the frequency range representing oscillations among the group of generators. Load characteristics, in particular, have a major effect on the stability of inter-area modes. The time frame of interest in small-signal stability studies is of the order of 10–20 s following a disturbance.

As discussed in Sect. 6.2.1, the DFIG-based design consisting of the power electronics converter imparts significant impact on the dynamic performance of the system. The important point to note is that the DFIGs are asynchronous machines, but the machines inject power into the system and as a result will affect the angular positions of all the other synchronous generators. In addition, DFIGs do not contribute to the inertia in the system. Hence, the synchronizing capability of the synchronous generators in the system will be affected. This leads to a significant impact on the electromechanical modes of oscillation. The following section discusses how change in system parameters can affect the electromechanical modes of oscillations of the system.

6.2.3 Eigenvalue Sensitivity

The effect of system parameters on overall system dynamics can be examined by evaluating the sensitivity of the eigenvalues with respect to variations in system parameters [11–14]. This section presents the mathematical derivation of the sensitivity of eigenvalue with respect to the system parameter x_j.

Taking the partial derivative of (5.10) with respect to x_j on both sides,

$$\frac{\partial A}{\partial x_j} v_i + A \frac{\partial v_i}{\partial x_j} = \lambda_i \frac{\partial v_i}{\partial x_j} + \frac{\partial \lambda_i}{\partial x_j}_j v_i \qquad (6.13)$$

where A is an $n \times n$ state matrix, λ_i is the ith eigenvalue, v_i is an $n \times 1$ right eigenvector of A corresponding to λ_i and n is the order of the system.

Premultiplying (6.13) by w_i yields,

$$w_i \frac{\partial A}{\partial x_j} v_i + w_i A \frac{\partial v_i}{\partial x_j} = w_i \lambda_i \frac{\partial v_i}{\partial x_j} + w_i \frac{\partial \lambda_i}{\partial x_j} v_i \qquad (6.14)$$

where w_i is the $1 \times n$ left eigenvector of A corresponding to λ_i.

Rearranging (6.14),

$$w_i \frac{\partial A}{\partial x_j} v_i + w_i (A - \lambda_i I) \frac{\partial v_i}{\partial x_j} = w_i \frac{\partial \lambda_i}{\partial x_j} v_i. \qquad (6.15)$$

Again,

$$w_i (A - \lambda_i I) = 0. \qquad (6.16)$$

Substituting (6.16) into (6.15) yields

$$w_i \frac{\partial A}{\partial x_j} v_i = w_i \frac{\partial \lambda_i}{\partial x_j} v_i. \qquad (6.17)$$

Rearranging (6.17) yields

$$\frac{\partial \lambda_i}{\partial x_j} = \frac{w_i \frac{\partial A}{\partial x_j} v_i}{w_i v_i}. \qquad (6.18)$$

Thus, sensitivity of the eigenvalue λ_i to the system parameter x_j is given by (6.18). This expression will be used to further explore the impact of the change of parameters due to the introduction of DFIGs on the modes of oscillation of the system.

The guiding principle for examining the impact of increased penetration of DFIG-based wind farms on small-signal stability is based on the premise that with the penetration of DFIG-based wind farms the effective inertia of the system will be reduced. As a result, a critical step in the examination of system behavior with increased DFIG penetration is to identify how the small-signal stability behavior changes with change in inertia. A logical progression of this chain of thought leads to the evaluation of eigenvalue sensitivity with respect to generator inertia. From (6.18), the eigenvalue sensitivity with respect to inertia can be evaluated as

$$\frac{\partial \lambda_i}{\partial H_j} = \frac{w_i \frac{\partial A}{\partial H_j} v_i}{w_i v_j} \qquad (6.19)$$

where H_j is the inertia of jth conventional synchronous generator, λ_i is the ith eigenvalue w_i and v_i is the left and right eigenvector corresponding to ith eigenvalue, respectively.

The key to the proposed analysis is to examine the sensitivity with respect to inertia and identify which modes of oscillation are affected in a detrimental fashion and which modes are affected in a beneficial fashion by the increased DFIG penetration.

An important point to note is that with increased penetration of DFIG, the synchronizing capability of the synchronous generators in the system will be affected. This leads to a significant impact on the electromechanical modes of oscillation. The proposed method provides a means of quantifying the effect of reduced inertia on the damping ratio of various modes of oscillation, as documented in [15, 16].

The following steps are followed to evaluate system response with respect to small disturbances:

- Replace all the DFIGs with conventional synchronous generators of the same MVA rating, which will represent the base case operating scenario for the assessment.
- Perform eigenvalue analysis in the frequency range: 0.1–2 Hz and damping ratio below a certain chosen threshold.
- Evaluate the sensitivity of the eigenvalues with respect to inertia (H_j) of each wind farm represented as a conventional synchronous machine, which is aimed at observing the effect of generator inertia on dynamic performance.
- Perform eigenvalue analysis for the case after introducing the existing, as well as planned DFIG wind farms in the system.

6.2.4 Example Study of Impact on Small-Signal Rotor Angle Stability

An example application of the approach described above is presented on a realistic test system. The study is carried out on a large system having over 22,000 buses, 3,104 generators with a total generation of 580,611 MW. All the modeling details provided within the base set of data are retained and represented in the analysis. This includes the modeling of governors on conventional generators. Within this large system, the study area that is experiencing a large increase of wind penetration has a total installed capacity of 4730.91 MW. The analysis is based on the information provided with regard to existing and planned increases in wind penetration. A total of 14 wind farms ranging from 14 to 200 MW with a total installed capacity of 1,460 MW are modeled as DFIGs. This information is used to set up the changed cases from the base case provided.

The system has transmission voltage levels, ranging from 34.5, 69, 161 to 345 kV. Wind turbine generators are connected to the grid at the 69 kV level. All new wind farms are represented using the GE 1.5 MW DFIG model available in a number of commercial software packages. Each DFIG unit has a rated power of 1.5 MW and MVA rating of 1.67 MVA. The reactive power capability of each unit corresponds to the power factor in the range +0.95/–0.90, which corresponds to $Q_{max} = 0.49$ MVAr and $Q_{min} = -0.73$ MVAr.

In the example presented on the test system considered, four different cases are analyzed. These include:

- Case A constitutes the case wherein all the existing DFIGs in the study area in the original base case are replaced by conventional round rotor synchronous machines of equivalent MVA rating.
- The original base case with existing DFIGs in the system is referred to as Case B.
- Case C constitutes the case wherein the penetration of DFIG-based WTGs in the study area is increased by 915 MW. The load in the study area is increased by 2 % (predicted load growth) and rest of the generation increase is exported to a designated nearby area.
- Case D constitutes the case wherein the DFIG wind farms with increased wind penetration are replaced by round rotor synchronous machines of corresponding MVA rating. Thus, in Case D, the round rotor machines representing the wind turbine generators are of higher MVA rating than in Case A.

The basis of this study lies in the premise that with the penetration of DFIG-based wind farms, effective inertia of the system will be reduced. Hence, as a first step toward studying the system behavior with increased DFIG penetration and to evaluate how the eigenvalues respond to change in inertia, a sensitivity analysis is carried out with respect to generator inertia. The eigenvalue sensitivity with respect to a specific system parameter indicates the degree to which a change in a particular parameter affects the eigenvalues. The sensitivity analysis with respect to inertia is only conducted for Case A, where all machines in the system are represented by conventional synchronous generators. The sensitivity of a given mode with respect to inertia of each wind farm replaced by a conventional synchronous generator is obtained. In this analysis, inertia was chosen to be the sensitivity parameter. The analysis inherently accounts for the insertion points of the wind farms in the system.

The analysis is carried out only for Case A, for all modes in a range of frequencies from 0.1 to 2 Hz. As the stability of a mode is determined by the real part of eigenvalue, the sensitivity of the real part is examined, and the mode which has the largest real part sensitivity to change in inertia is identified. Among the several modes of oscillation analyzed, the result of sensitivity analysis associated with the mode having significant detrimental real part sensitivity, in comparison to the real part of the eigenvalue as shown in Table 6.1.

The corresponding sensitivity values for the real part of this mode with respect to each of the 14 wind farm generators replaced by conventional synchronous machines of the same MVA rating are shown in Table 6.2. The real part sensitivities all having negative values as shown in Table 6.2 reveal that with the decrease in inertia at these locations, the eigenvalue will move toward the positive right half plane making the system less stable.

The next step in the analysis is to observe if the penetration of DFIGs has beneficial impact in terms of damping power system oscillations. The sensitivity with respect to inertia is examined for positive real part sensitivity. This identifies the mode where the increased penetration of DFIGs in the system results in

Table 6.1 Dominant mode with detrimental effect on damping

Real part (1/s)	Imaginary part (rad/s)	Frequency (Hz)	Damping ratio (%)
−0.0643	3.5177	0.5599	1.83

Table 6.2 Eigenvalue sensitivity corresponding to the dominant mode with detrimental effect on damping

No.	Generator bus #	Base value of inertia (s)	Sensitivity of real part $(1/s^2)$
1	32672	2.627	−0.0777
2	32644	5.7334	−0.0355
3	32702	3	−0.0679
4	32723	5.548	−0.0367
5	49045	5.2	−0.0383
6	49050	4.6	−0.0444
7	49075	4.2	−0.0475
8	52001	5.2039	−0.0389
9	55612	3.46	−0.0581
10	55678	4.3	−0.0467
11	55881	4	−0.0506
12	55891	4.418	−0.0466
13	55890	5.43	−0.037
14	55889	5.43	−0.0374

Table 6.3 Dominant mode with beneficial effect on damping

Real part (1/s)	Imaginary part (rad/s)	Frequency (Hz)	Damping ratio (%)
−0.0651	2.8291	0.4503	2.3

shifting the eigenvalues of the system state matrix towards the negative half plane. Among the several modes of oscillation analyzed, the result of sensitivity analysis associated with the mode having the largest positive real part sensitivity is shown in Table 6.3.

The corresponding real part sensitivity values for each of the 14 wind farm generators represented by conventional synchronous machines are shown in Table 6.4. The real part sensitivities are all positive in sign indicating that with the decrease in inertia at each of these locations the mode will move further into the left half plane making the system more stable.

Detailed eigenvalue analysis is conducted for Cases A to D in the frequency range of 0.1–2 Hz to substantiate the results of the sensitivity analysis. Table 6.5 shows the result of eigenvalue analysis corresponding to the mode listed in Table 6.1 which has detrimental impact with increased DFIG penetration.The critical mode is observed in all the four cases. The frequency of the mode is relatively unchanged. It is observed that the damping ratio associated with the mode is reduced from 1.83 % in Case A to 1.16 % in Case B. The damping ratio

Table 6.4 Eigenvalue sensitivity corresponding to the dominant mode with beneficial effect on damping

No.	Generator bus #	Base value of inertia (s)	Sensitivity of real part ($1/s^2$)
1	32672	2.627	0.0169
2	32644	5.7334	0.0078
3	32702	3	0.015
4	32723	5.548	0.008
5	49045	5.2	0.0075
6	49050	4.6	0.0092
7	49075	4.2	0.0104
8	52001	5.2039	0.0079
9	55612	3.46	0.0125
10	55678	4.3	0.0098
11	55881	4	0.0107
12	55891	4.418	0.0095
13	55890	5.43	0.0082
14	55889	5.43	0.008

Table 6.5 Result summary for Cases A, B, C, and D for dominant mode with detrimental effect on damping

Case	Real part (1/s)	Imaginary part (rad/s)	Frequency (Hz)	Damping ratio (%)
A	−0.0643	3.5177	0.5599	1.83
B	−0.0412	3.5516	0.5653	1.16
C	−0.0239	3.5238	0.5608	0.68
D	−0.0427	3.4948	0.5562	1.22

Table 6.6 Result summary for Cases A, B, C, and D for the dominant mode with beneficial effect on damping

Case	Real (1/s)	Imaginary (rad/s)	Frequency (Hz)	Damping ratio (%)
A	−0.0651	2.8291	0.4503	2.3
B	−0.0725	2.8399	0.452	2.55
C	−0.0756	2.8189	0.4486	2.68
D	−0.0566	2.805	0.4464	2.02

has further reduced to 0.68 % in the Case C. However, in Case D with DFIGs replaced by conventional machines the damping ratio is improved to 1.22 %.

The change in damping ratio due to increased penetration of DFIGs accurately reflects the trend indicated by sensitivity analysis. In Cases B and C where the inertia is reduced due to the inclusions of DFIGs, the damping has dropped. In Cases A and D where the machines are represented as conventional synchronous machines of equivalent rating, the damping is higher. The eigenvalue analysis carried out for the mode with beneficial eigenvalue sensitivity is shown in Table 6.6.

The damping ratio associated with the mode having a beneficial impact has increased from 2.3 % in Case A to 2.55 % in Case B. It is also observed that the damping ratio has further improved to 2.68 % in Case C. However, in Case D, DFIGs replaced by conventional machines, the damping ratio is reduced to 2.02 %. These results again reflect the trend in damping change identified by the sensitivity analysis and also show the small change in damping as reflected by the sensitivity values in comparison to the mode that was detrimentally affected by a change in inertia.

These results reveal that eigenvalue sensitivity with respect to inertia provides an effective measure of evaluating impact of DFIG penetration on system dynamic performance. The detailed eigenvalue analysis carried out for each of the four cases is found to substantiate the results obtained from sensitivity analysis. This method identifies both the beneficial as well as detrimental impact due to increased DFIG penetration.

6.2.5 Impact on Transient Rotor Angle Stability

In examining transient rotor angle stability, the primary goal is to determine the ability of a power system to maintain synchronism when subjected to a large disturbance. In this situation, the complete nonlinear representation of the system model is utilized and the conventional time domain simulations are conducted using appropriate models for the various components. For wind generators, depending on their type, the models developed in Chap. 5 are utilized. These models together with those of the other components and the conventional synchronous generators are then formulated as a system of differential algebraic equations and then solved numerically for a given disturbance scenario. In the case of wind generators, the wind farm is typically modeled as an equivalent generator (detailed in Chap. 4) and the wind speed is kept constant. There are several commercial software packages that allow a wind speed model to examine the impact of wind farm output variation on the rest of the system.

Several recent efforts have examined the impact of DFIG wind turbine generators on transient stability [17–20]. In these efforts, the primary objective is to analyze the impact of the DFIG wind turbine generator-based wind farms on the transient stability of the rest of the system. In this regard, Fig. 6.1 can be again utilized to motivate the issue at hand. A wind farm will inject asynchronous power at the point of interconnection with the grid. This injection alters the flow in the rest of the system depending on the point of injection and the size of the injection. In addition, the rotor angle positions of the other synchronous generators in the system adjust to account for this flow. A large disturbance in the system will affect the various synchronous generators depending on the location and severity of the disturbance. Additionally, the imbalance resulting from the disturbance will also be affected by the location and size of the injections from the various wind farms. Detailed time domain simulations will need to be conducted at varying operating conditions and at different levels of wind farm outputs to examine the impact of wind penetration on the transient rotor angle stability of the other synchronous generators in the system.

As observed in small-signal stability behavior, depending on the disturbance and injections from the wind farms, certain generators would be either detrimentally or beneficially affected. An example of this behavior is depicted in the following section.

6.2.6 Example Study of Impact on Transient Rotor Angle Stability

Fault scenario to depict detrimental impact of wind penetration

The two modes identified in Sect. 6.2.4 as being detrimentally and beneficially affected are excited by applying large disturbances. The plots of the dominant generators are shown to examine the impact of increased wind penetration on large disturbances. The objective here is to observe whether a large disturbance will excite the mode shown in Table 6.1. A three-phase fault is applied at a particular bus in the system, which is a 230 kV bus and is cleared after five cycles.

As the dominant mode considered from small-signal analysis corresponds to the speed state of the machines, the generator speed is observed in time domain. Figure 6.2 shows the generator speed corresponding to a generator which is participating in the mode shown in Table 6.1. Observing the oscillation corresponding to the last swing of Fig. 6.2, the least amount of damping is found in Case C, followed by Case B, and Case D. The damping is found to be the highest in Case A. The damping behavior observed in time domain corresponds exactly to the damping ratio provided by the eigenvalue analysis in Table 6.5. This fault scenario is thus found to have detrimental impact on system performance with respect to increased wind penetration.

Fault scenario to examine beneficial impact of wind penetration

Following the same procedure adopted for the detrimental case for exciting the required mode following a fault, a three-phase fault applied at another 230 kV is cleared after 6.6 cycles. This is then found to excite the mode depicted in Table 6.3. As the dominant mode considered for small-signal analysis corresponds to the speed state of the machines, the generator speeds are observed in time domain.

Figure 6.3 shows the generator speed corresponding to the dominant generator for the mode shown in Table 6.3. The results show that the system is transiently stable for all the four cases and confirms the mode damping ratio results depicted in Table 6.6.

The damping values in the plots for Cases A–C are very close to each other and accurately reflect the damping ratio results shown in Table 6.6. For Case D the damping is markedly lower and verifies that the higher inertia case will have the lower damping. The oscillation damping is increased with an increase in DFIG penetration. This fault scenario is thus found to have a beneficial impact with respect to increased wind penetration. The analyses clearly indicate that the results

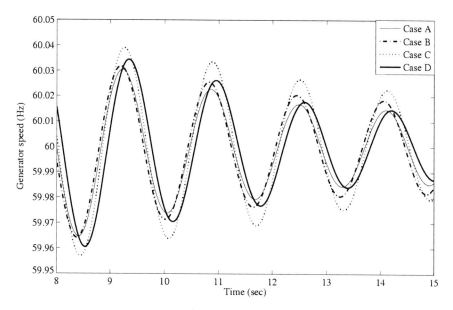

Fig. 6.2 Generator speed for dominant generator in detrimentally affected mode for Cases A, B, C, and D

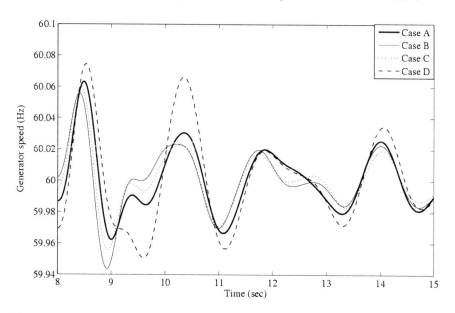

Fig. 6.3 Generator speed for dominant generator in beneficially affected mode for Cases A, B, C, and D

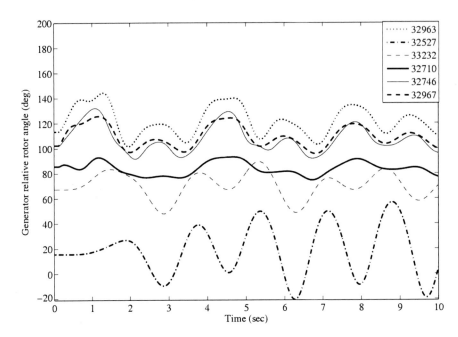

Fig. 6.4 Generator relative rotor angle for Case B

obtained from sensitivity analysis and from eigenvalue analysis are confirmed by exciting the mode in time domain.

Fault scenario to examine impact of increased wind penetration

The objective here is to observe whether a large disturbance will affect the transient stability performance for an increased penetration of wind generation. A three-phase fault is applied at bus 32946 in the system described in Sect. 6.2.4, which is at 345 kV bus. The fault is cleared after 4.5 cycles followed by clearing the line connecting the buses 32969 and 32946.

The time domain simulation is carried out to observe the effect of increased DFIG penetration on the system. The relative rotor angle plots for Cases B and C are shown in Figs. 6.4, 6.5, 6.6.

The system is found to be transiently secure in Case B, whereas it is found to be transiently insecure in Case C. The machines swinging apart in Case C as a result of the fault can be segregated into accelerating and decelerating groups. The machines that accelerate are shown in Fig. 6.5 and the machines that decelerate are shown in Fig. 6.6. The result of the disturbance is a large inter-area phenomenon in the case with increased wind penetration.

These cases illustrate the point made earlier in this section that depending on the size and location of the wind power injection, and the location and severity of the disturbance, transient stability could be affected either detrimentally, or beneficially.

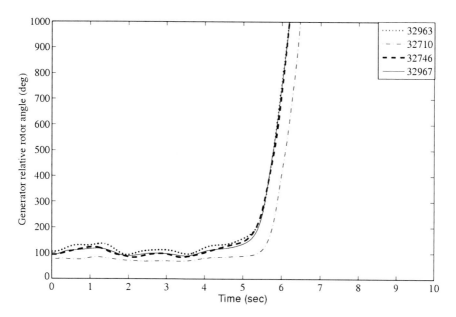

Fig. 6.5 Generator relative rotor angle for machines accelerating in Case C

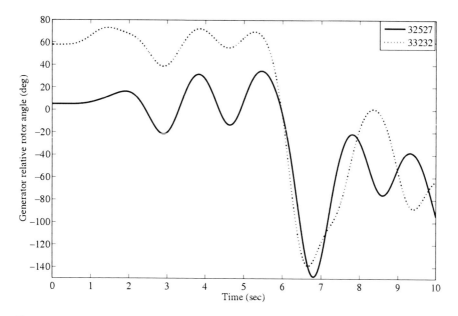

Fig. 6.6 Generator relative rotor angle for machines decelerating in Case C

6.3 Impact on Voltage Response and Stability

In this section we will examine the impact of DFIG-based wind turbine generators on voltage response following disturbances in the vicinity of the wind farm and on voltage stability of the system to which they are interconnected. In examining these aspects of voltage response and stability, the reactive power control capability of the DFIG-based wind turbine generator and of the wind farm as a whole becomes important. As a result, an analysis of the reactive power capability of the DFIG-based wind turbine generator and of the wind farm as a whole is required. Several recent publications have examined this aspect of voltage response and stability performance [21–27]. The critical aspects related to voltage response and stability and the associated reactive capability and control of DFIG-based wind farms will be discussed.

6.3.1 Operating Modes of a DFIG Wind Turbine Generator

The DFIG wind turbine rotor side converter has the capability to adjust the rotor currents to obtain the desired real and reactive power outputs on the stator side. The details of these aspects were provided in Chap. 5. The real power set point is obtained using a maximum power tracking scheme as discussed in Chap. 5. The reactive power set point is dependent on the control mode of the DFIG. The two commonly used control modes are:

1. Power factor control
2. Voltage control

The stator real and reactive powers are controlled in the power factor control mode to maintain a constant power factor at the point of interconnection. The reactive power is controlled in the voltage control mode to maintain the voltage magnitude at a specified value in the voltage control mode. The stator side converter is normally set at unity power factor. Keeping these facts in mind, important aspects of voltage response and voltage stability will be examined.

6.3.2 Voltage Ride Through

Historically, wind turbines were connected to distribution systems. In the event of a large disturbance resulting in a significant dip in voltage close to the wind turbine it was accepted practice to trip the wind turbine. This action was based on the reasoning that distribution systems being supplied by radial systems could result in the wind farm being connected to the system following the disturbance and result in an islanded operation, which could cause unacceptable overvoltage conditions [28]. With the evolution of wind turbine technology and the

Fig. 6.7 Pictorial
representation of FERC
Order 661-A

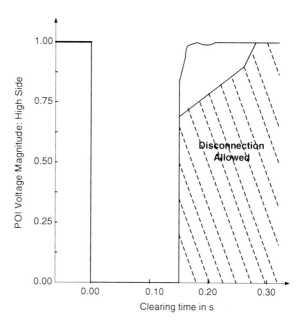

development of large wind warms connected to the bulk transmission network, the issue of voltage ride through sometimes also referred to as low voltage ride through or fault ride through, became an important operating issue. This led to the development of voltage ride through requirements in various parts of the world. In the US the Federal Energy Regulatory Commission (FERC) mandated Order 661-A, which requires wind farms to remain online in the presence of severe voltage disturbances for a specified period of time and with a defined voltage profile [29].

The regulations require wind plants to remain in service during three-phase faults with normal clearing (this is a time period of approximately 4–9 cycles) and for single line to ground faults with delayed clearing. The regulations also require the subsequent postfault voltage to return to the prefault voltage level unless the clearing of the fault effectively disconnects the wind generator from the system. In the event that a fault persists for a period greater than the clearing duration specified, or the postfault voltage does not recover to the specified value, the wind park may disconnect from the transmission system. Wind generating plants must remain connected during such faults with voltage levels down to zero volts as measured at the high side of the point of interconnection (POI) transformer. Figure 6.7 pictorially depicts the fault ride through regulation.

Given this regulatory requirements on fault ride through performance, the primary issue associated with DFIG machines is the current sensitivity of the insulated bridge bipolar transistors (IGBTs) that make up the power electronics converter [21]. These devices are subject to damage if the converter current limitations are exceeded. The rotor side converter of the DFIG is connected to the rotor via slip rings as discussed in earlier chapters in this book. A solution to isolate the rotor side converter and protect it consists of using a crowbar circuit (described in Chap. 3), which could be short circuited electrically parallel to the rotor windings. This action electrically isolates the rotor side converter from damaging transient currents that could be induced in the rotor winding from the stator side of the machine during disturbances.

6.3.3 Power Capability Curve of a DFIG Machine

Conventional electromechanical machines have inherent limitations that limit the power production. In order to determine the true power production capability of the machine it is important to know the operating characteristics of the machine. An excellent discussion of power capability curve characteristics for a synchronous machine is provided in [28, Sect. 5.4]. In the case of a DFIG, two factors influence the constraints that define the electrical power capability: (a) the wound rotor induction machine and (b) the power electronics associated with the converters. Another significant distinction between a DFIG wind turbine generator and a conventional synchronous generator is that a DFIG power output is dependent on wind speed and the maximum power tracking characteristic of the machine as described in Chap. 1. As a result, the real power limits are set by wind speed. On the other hand, the reactive power capability is determined by machine design, which sets limitations on the stator and rotor circuits in terms of the applied currents and voltages. Hence, the maximum power capability is dependent on the design of the machine in conjunction with the back-to-back converter.

An approach to developing the capability curve for a DFIG wind farm with maximum power tracking capability for each wind turbine in the farm is developed in [22]. Based on this approach, [23] have developed the power capability curve of a single DFIG wind turbine generator and assume that the capability of the single unit could be scaled up to accurately aggregate the behavior of the wind farm. Figure 6.8 shows the DFIG wind farm static power capability curve in p.u. from [23].

A discussion of the underlying details that justify the use of a scaled capability curve to represent the capability of wind farm is important in examining the use of the equivalent capability curve in system studies. In most wind farms, all the turbines are typically of the same model. The reactive capability of the farm is given by the product of the reactive capability of each turbine and the total number of turbines in the farm. Each farm also has a supervisory controller that determines this plant's rating, typically in kVAr. The supervisory controller measures the voltage magnitude and the reactive power delivered at the POI of the wind farm. The reactive power delivered to the POI is not the same as the product of the reactive power output of each individual turbine and the number of turbines in the wind farm because there are reactive power losses. The supervisory control then produces a reactive power command signal in kVAr, which is not allowed to exceed the wind farm rating. The reactive power command is divided by the total number of wind turbines in the farm and distributed to each machine. There are a few other nuances such as when the supervisory control also has

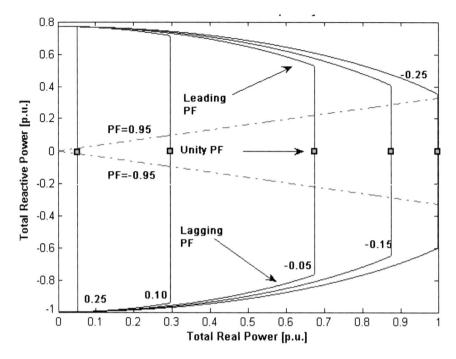

Fig. 6.8 Static power capability curve of DFIG wind farm in p.u. [23]

to deal with shunt capacitors or reactors. The supervisory control also knows when some individual machines are off-line and therefore compensates accordingly. In the model used in studies, the supervisory controller and each wind turbine are expressed in per unit, like all other models. Since the wind farm is represented by an equivalent machine whose rating is determined by the product of the rating of each individual turbine and the number of turbines in the wind farm, the base for per unit calculation is consistent with the equivalent representation to each individual turbine.

6.3.4 Impact of DFIG Wind Turbines on Steady-State Voltage Stability

Doubly fed induction generators wind turbine generators have the inherent ability to control reactive power output without the need for additional reactive power support. The impact of DFIG wind turbine generators on steady-state voltage stability will depend on the location of the wind farm interconnection to the grid and the location of areas in the grid that are susceptible to voltage problems. These areas may be prone to voltage problems due to lack of reactive resources, the nature of the load in that area or due to the capability of the transmission system in that area. In many parts of the world, wind farms are typically connected to either the distribution system or to the subtransmission (69 kV). As a result, there is impact on the bulk transmission system in terms of providing voltage support, which should be carefully investigated.

A DFIG wind turbine generator is capable of operation in one of two control modes: (a) fixed power factor control and (b) terminal voltage control. In the case of power factor control, the reactive power production is controlled to obtain a specified power factor. In the case of terminal voltage control, the reactive power is controlled to meet a target voltage at a specified bus.

Various approaches have been proposed to implement terminal voltage control [23, 25–27]. These efforts have shown that enhanced voltage control improves voltage performance in the system. In analyzing the impact of DFIG wind turbines on steady-state voltage stability characteristics an important issue to capture is the variable nature of the wind. This variability does have an impact on the ability of the wind farm to provide the needed reactive power support. There are two approaches to capture the variability of the wind resource. One is by using a probabilistic power flow approach and the other by using a detailed times series power flow approach.

In the probabilistic power flow approach, historical data is typically used to statistically characterize wind variation in terms of a probability density function. Care should be taken, however, to ascertain that sufficient historical data is collected at all the wind farm sites since the spatio-temporal variability of the wind also needs to be captured. The probabilistic power flow analysis will provide important statistical measures of the system performance, but will not capture the detailed mechanism of the voltage related problems. In comparison, a time-series power flow simulation using time-series data for wind speeds and load variation can be effectively used to construct a sequential simulation around conditions of interest at different points of the annual operating horizon to examine the ability of wind farms to provide voltage support. The time-series data can also be used to deterministically model the worst-case point within the data set. It is important to note that, system loading and wind power output data should be carefully examined to ensure that the chosen time-series sequence captures the complex relationship between wind and load. In systems with high wind penetration, the worst-case operating conditions with regard to voltage stability typically occur when the wind generation serves a significant portion of the load demand and the voltage support mechanisms are inadequate. Another equally important aspect is to recognize that the terminal bus voltage control capability of DFIG wind turbine generators is highly localized. The terminal voltage control is capable of significantly impacting only the portion of the system in which the wind generation is located. Hence, it is important to represent region-specific wind power data for different portions of the system.

A detailed development of the time-series power flow-based analysis is given in [24]. In this paper, a detailed demonstration of the approach on the Irish system in which significant wind penetration occurs at both the transmission and distribution levels is analyzed. The results shown in the paper demonstrate that the voltage control features of the DFIG wind turbines are capable of improving the steady-state voltage stability margin at both the distribution and transmission level buses in the system.

6.4 Impact of DFIG Wind Turbine Generators on System Frequency Response

In power systems, frequency is controlled by balancing power generation against load demand on a second-by-second basis. There is a need for continuous adjustment of generator output as the load demand varies. At the same time, the system should be able to respond to occasional larger mismatches in generation and load, caused, for example, by the tripping of a large generator or a large load.

Whenever there is load/generation imbalance, synchronous generators in a system respond in three stages to bring the system back to normal operation. The initial stage is characterized by the release or absorption of kinetic energy of the rotating mass —this is a physical and inherent characteristic of synchronous generators. For example, if there is sudden increase in load, electrical torque increases to supply load increase, while mechanical torque of the turbine initially remains constant. Hence, the synchronous machines decelerate. The turbine-generator decelerates and rotor speed drops, as kinetic energy is released to supply the load change. This response is called "inertial response." As the frequency deviation exceeds certain limits, the turbine-governor control will be activated to change the power input to the prime mover. The rotor acceleration eventually becomes zero and the frequency reaches a new steady state. This is referred to as "primary frequency control." After primary frequency support, there still exists steady-state frequency error. To correct this error, the governor set points are changed and the frequency is brought back to nominal value. This is called "secondary frequency control." These three phenomena take place in succession in any system to restore the normal operating equilibrium. A detailed analytical treatment of these three stages of evolution in the frequency response following a power impact assuming a classical machine model is developed in [4, Chap. 3]. The analytical development demonstrates that during the inertial response, the power impact is shared among the synchronous generators based on their electrical distance from the point of the power impact. This translates to the machines sharing the power impact in the ratio of their synchronizing power coefficient. As the transient progresses, the impact is then shared among the machines in the ratio of their inertias. It is after this period that the impact is then shared among the machines based on their governor settings and corresponds to the primary frequency response. Any deviation from the nominal frequency is then corrected by the secondary frequency control.

The dynamic performance of DFIG units as seen by the grid is completely governed by the power electronic converters that control them. With conventional control, where rotor currents are always controlled to extract maximum energy from the wind by varying the rotor speed, the inertia of the turbine is effectively decoupled from the system. With the penetration of DFIG-based wind farms, the effective inertia of the system will be reduced. With the inertia decreasing due to a large number of DFIG wind turbines in operation, system reliability following large disturbances could be significantly affected. An excellent discussion of this impact on system frequency is presented in [30]. Since the inertia of the DFIG machines are completely masked from the network due to back-to-back converters, the frequency response especially in terms of the frequency nadir following loss of generation could be reduced if DFIG wind generators displace a considerable amount of conventional synchronous generators. A detailed analysis of this specific aspect is presented in [2, 30]. A key point to note is that the nadir in the frequency response could be affected greatly by the amount of spinning reserve available. If sufficient reserve is maintained the frequency nadir can be controlled. This aspect is addressed in [1, 30]. An approach that could be considered is the emulation of inertial response in DFIG wind generators using power electronic converters. This concept has been proposed in [31–34]. Details of the approach developed in [34] are presented in Sect. 6.4.1.

6.4.1 Frequency Support from a DFIG Wind Turbine

The approach developed in [34] to emulate inertia in DFIG wind turbines in order to provide frequency support is detailed. The strategy is based on the idea of changing the torque set point of the DFIG for changes in grid frequency. The schematic of the supplementary control proposed in the present work is shown in Fig. 6.9. From Fig. 6.10 (a modification of Fig. 4.3 to illustrate the function of the supplementary inertia controller) and Fig. 6.9, it is noted that the supplementary control signal (ΔT_{set}) is algebraically added to the torque set point (T_{set}). The torque set point increases whenever grid frequency drops from the nominal value of 1 p.u.

The input to the supplementary controller is the frequency deviation at the POI. The study is carried out in a system consisting of several wind farms dispersed geographically. During the first few seconds, even the same disturbance can have wide ranges of impact on the machines involved. Hence, transiently, the frequency deviation might vary at different buses in the system. As the controller is required to respond accordingly, the deviation of grid frequency is used as a control input. In addition, the support from the DFIGs is dependent on the kinetic energy stored in the turbine blades. This support is limited by the inertia constant of the turbines, which in turn depends on its MVA rating. Hence, the gain of the supplementary control block is adjusted based on the frequency deviation and MVA rating of the wind farm. The gain is evaluated as follows:

$$G = \frac{1}{\Delta\omega_{max}} \frac{S_{BM}}{S_{B,sys}} \qquad (6.20)$$

where G is the gain of the supplementary control loop, S_{BM} is the rated MVA of the DFIG wind farm, $S_{B,sys}$ is the system MVA base, and $\Delta\omega_{max}$ is the maximum deviation in grid frequency at the POI among all the wind farms. The maximum deviation in grid frequency is evaluated by simulating a series of large disturbances in terms of generation drops that the system can withstand without being transiently unstable.

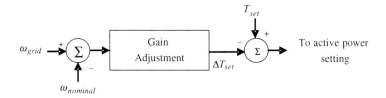

Fig. 6.9   Supplementary inertial control loop

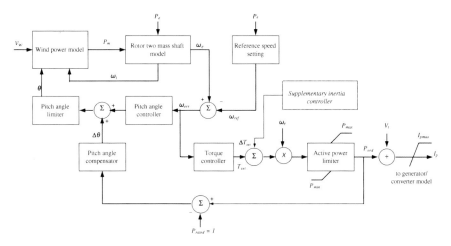

Fig. 6.10   Schematic of DFIG controllers with the supplementary inertia controller

When the objective is to provide inertial support, the time frame of interest is the first few seconds after the disturbance. A DFIGs equipped with the supplementary control loop can support the system during this period. The additional electrical power output is provided via a concomitant reduction in rotor speed. Furthermore, as the primary input, the wind flow cannot be changed; kinetic energy stored in the rotating mass of a wind turbine is assumed to provide primary frequency control. This, however, depends on factors such as rating and margin on the converter and the aerodynamics of the turbine. Due to limited stored kinetic energy, the support can only be provided for a short period of time. The final steady-state frequency error, if it is nonzero, should be adjusted by secondary frequency control as discussed in Sect. 6.4. As the supplementary control is not aimed at correcting the steady-state error, proportional gain is sufficient.

6.4.2 Pitch Compensation Adjustment

Owing to variable speed operation, the stored kinetic energy and, accordingly, the inertial response of the DFIG is dependent on the captured mechanical power. The latter is governed by the pitch angle controller. Maximum benefit can be achieved when the pitch angle controller aids the operation of the supplementary frequency controller. Conventional parameters of pitch angle controller can be tailored so that mechanical power can be changed, thus providing primary frequency response under transient condition. This is done by suitably varying the PI controller gain of the pitch compensator. The parameters should be varied in such a way that the pitch compensator does not act to increase the pitch angle during the transient period when the system is subjected to loss of generation.

A heuristic approach based on trial and error is used. The "cause and effect" approach is adopted when the controller is tuned. The parameters are selected such that the pitch compensator reduces the pitch angle, thus avoiding the transient drop in extracted mechanical power during the condition when the system demands more active power. This action aids the purpose of the supplementary controller, thus increasing the power output during the transient with response to drop in grid frequency.

6.4.3 Maximum Power Order Adjustment

The additional power supplied by the DFIG depends on P_{max}, which in turn depends on factors such as the rating and margin on the converter and the aerodynamics of the turbine. The idea here is to increase the value of P_{max} such that inertial response of DFIG is improved during the transient event. To be within the converter design limits, the increased value of P_{max} should not demand the current command beyond the short-term current capability of the converter (I_{pmax}).

6.4.4 Example of Effectiveness of Supplementary Inertia Control

The same test system utilized in Sect. 6.2.4 is used to examine the effectiveness of the supplementary inertia control. In order to create load/generation imbalance, four generators in the study area, totaling 1,950 MW, are disconnected at $t = 1$ s. In order to make the frequency drop more apparent in the system, three 345 kV lines are disconnected at the same time. These are the lines transmitting power from generators in the western region to the load centers in the central region of the study area. These changes result in a significant dearth of generation in the load centers and hence result in frequency deviations in a system that is quite robust.

Furthermore, the inertial response of the DFIG is dependent on the initial operating condition. Since this is a variable speed machine, the stored kinetic energy is different at different operating conditions. At low load some wind turbine generator units may be off-line reducing the total wind farm output. This is different in comparison with synchronous generation where the inertial response is always constant. In order to quantify maximum inertial response from the DFIG, the present study considers the units to be operating at rated wind speed.

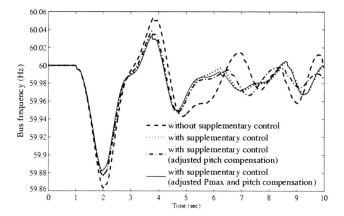

Fig. 6.11   Frequency at bus 32969 (345 kV) for Case B

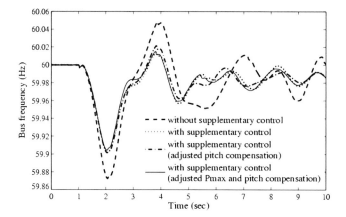

Fig. 6.12   Frequency at bus 32969 (345 kV) for Case C

The operating condition considered here is the summer peak load period where the wind farms are providing maximum power output.

The wind farms are connected at the distribution voltage level as mentioned before and spread from the eastern to western part of the system. The distribution voltage level is stepped up and power from all the wind farms is transmitted through a single 161 kV line. This line feeds the load at the central part of the system, which had also been fed by the generators being disconnected as mentioned earlier when the example was introduced in Sect. 6.4.4.

Time domain simulation is carried out for the contingency identified above for the case with and without the supplementary controller. The cases, designated as Case B and Case C in Sect. 5.2.4, are considered. A system bus (bus 32969) at the 345 kV voltage level with the highest frequency deviation and two other 161 kV buses (bus 24222 and 32729), which are electrically closer to the wind farm with

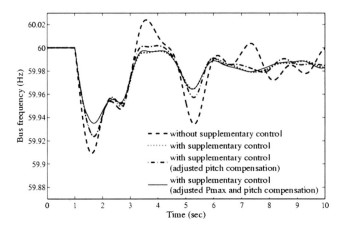

Fig. 6.13   Frequency at bus 24222 (161 kV) for Case B

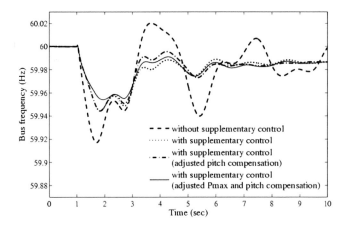

Fig. 6.14   Frequency at bus 24222 (161 kV) for Case C

the highest frequency deviation, are depicted in the figures shown below (Figs. 6.11, 6.12, 6.13, 6.14, 6.15, 6.16).

These plots clearly illustrate the efficacy of the supplementary controller in providing frequency support, and as a result, significantly improved frequency responses are observed. The effectiveness of the supplementary controller in providing inertial support is also examined in a little more detail. For the disturbance considered, the electrical power output of a DFIG wind turbine generator at bus 32672 is shown in Fig. 6.17. It is observed that the supplementary control significantly improves the power output characteristics and enables the DFIG to ramp up its output in response to a large loss of generation disturbance.

The corresponding DFIG rotor speed is shown in Fig. 6.18. It is seen that the increase in power output seen in Fig. 6.17 is accompanied by a corresponding drop

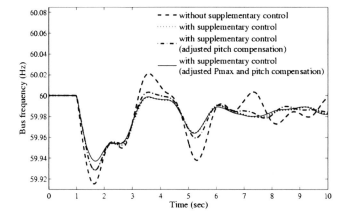

Fig. 6.15   Frequency at bus 32729 (69 kV) for Case B

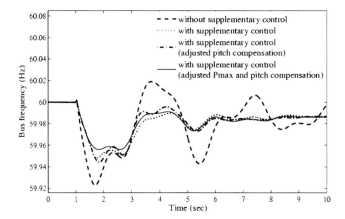

Fig. 6.16   Frequency at bus 32729 (69 kV) for Case C

in rotor speed and emulates the inertial response of a synchronous generator. It should be noted, however, that the inertial support cannot be provided for a significant period of time and is not permanent. This can be observed in the power output behavior after the first peak. Following inertial response, in order to return to steady state, only a part of the aerodynamic power is transmitted to the grid and the rest of the power is used to speed up the turbine back to its optimal speed. Thus, the power output of the DFIG momentarily drops to a value lower than its initial (rated) value immediately after providing inertial support.

This completes the discussion of the impact of DFIG-based wind turbine generators on three key components of dynamic performance; rotor angle stability, voltage stability, and frequency stability. The critical issue related to each aspect of dynamic performance has been discussed and highlighted using simulated responses on actual systems with significant wind penetration.

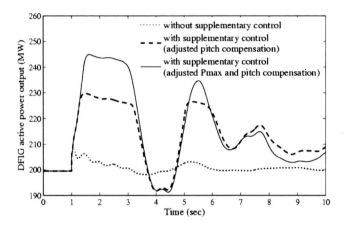

Fig. 6.17 Power output of DFIG at bus 32672 with and without supplementary control

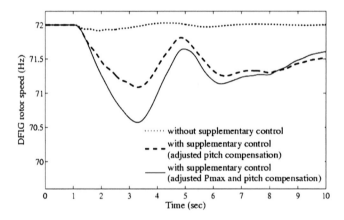

Fig. 6.18 Rotor speed of DFIG at bus 32672 with and without supplementary control

References

1. Lew D, Piwko R, Project Managers. Western wind and solar integration study. Subcontract report NREL/SR-550-47434, Prepared for NREL by GE Energy, May 2010
2. Vittal V, McCalley JD, Ajjarapu V, Shanbhag U (2009) Impact of increased DFIG wind penetration on power systems and markets. PSERC final report, Aug 2009
3. Kundur P, Paserba J, Ajjarapu V, Andersson G, Bose A, Canizares C, Hatziargyriou N, Hill D, Stankovic A, Taylor C, Van Cutsem T, Vittal V (2004) Definition and classification of power system stability, IEEE/CIGRE joint task force on stability terms and definitions report. IEEE Trans Power Sys 19(3):1387–1401
4. Anderson PM, Fouad AA (2003) Power system control and stability, 2nd edn. Wiley Interscience, Piscataway
5. Sanchez Gasca JJ, Miller NW, Price WW (2004) A modal analysis of a two-area system with significant wind power penetration. Proceedings of the 2004 IEEE PES power systems conference and exposition, pp 1148–1152

6. Slootweg JG, Kling WL (2003) The impact of large scale wind power generation on power system oscillations. Electr Power Sys Res 67(1):9–20
7. Fernandez RD, Mantz RJ, Battaiotto PE (2007) Impact of wind farms on power system—an eigenvalue analysis approach. Renew Energy 32(10):1676–1688
8. Xu F, Zhang XP, Godfrey K, Ju P (2007) Small-signal stability analysis and optimal control of a wind turbine with doubly fed induction generator. Proceedings of the 2007 IET generation, transmission & distribution, pp 751–760
9. Mendonca A, Lopes JAP (2005) Impact of large scale wind power integration on small-signal stability. Proceedings of the 2005 international conference on future power systems, pp 1–5
10. Mei F, Pal B (2007) Modal analysis of grid connected doubly fed induction generators. IEEE Trans Energy Convers 22(3):728–736
11. Faddeev DK, Faddeeva VN (1963) Computational methods of linear algebra. W. H. Freeman and Company, San Francisco and London
12. Van Ness JE, Boyle JM, Imad FP (1965) Sensitivities of large, multiple—loop control systems. IEEE Trans Autom Control 10(3):308–315
13. Smed T (1993) Feasible eigenvalue sensitivity for large power systems. IEEE Trans Power Sys 8(2):555–561
14. Ma J, Dong ZY, Zhang P (2006) Eigenvalue sensitivity analysis for dynamic power system. Proceedings of the 2006 international conference on power system technology, pp 1–7
15. Gautam D, Vittal V, Harbour T (2009) Impact of increased penetration of DFIG based wind turbine generators on transient and Small-signal stability of power systems. IEEE Trans Power Sys 24(3):1426–1434
16. Gautam D (2010) Impact of increased penetration of DFIG based wind turbine generator on rotor angle stability of power systems. Ph.D. Dissertation, Arizona State University
17. Kayikci M, Milanovic JV (2008) Assessing transient response of DFIG—based wind plants—the influence of model simplifications and parameters. IEEE Trans Power Sys 23(2):545–554
18. Nunes MVA, Lopes JAP, Zurn HH, Bezerra UH, Almeida RG (2004) Influence of the variable-speed wind generators in transient stability margin of the conventional generators integrated in electrical grids. IEEE Trans Energy Convers 19(4):692–701
19. Muljadi E, Butterfield CP, Parsons B, Ellis A (2008) Effect of variable speed wind turbine generator on stability of a weak grid. IEEE Trans Power Sys 22(1):29–35
20. Vittal E, O'Malley M, Keane A (2012) Rotor angle stability with high penetrations of wind generation. IEEE Trans Power Sys. 27(1):353–362
21. Hansen AD, Michalke G (2007) Fault ride-through capability of DFIG wind turbines. Renew Energy 32:1594–1610
22. Lund T, Sorensen P, Eek J (2007) Reactive power capability of a wind turbine with doubly fed induction generator. Wind Energy 10:379–394
23. Konopinski RJ, Vijayan P, Ajjarapu V (2009) Extended reactive capability of DFIG wind parks for enhanced system performance. IEEE Trans Power Sys 24(3):1346–1355
24. Vittal E, O'Malley M, Keane A (2010) A steady-state voltage stability analysis of power systems with high penetrations of wind. IEEE Trans Power Sys 25(1):433–442
25. Kayikci M, Milanovic JV (2007) Reactive power control strategies for DFIG-based plants. IEEE Trans Energy Convers 22(2):389–396
26. Tapia G, Tapia A, Ostolazam JX (2007) Proportional-integral regulator-based approach to wind farm reactive power management for secondary voltage control. IEEE Trans Energy Convers 22(2):488–498
27. Cartwright P, Holdsworth L, Ekanayake JB, Jenkins N (2004) Co-ordinated voltage control strategy for a doubly-fed induction generator (DFIG)-based wind farm. Proc Inst Electr Eng Gen Transm Distrib 151(4):495–502
28. Cigré Report 328 (2007) Modeling and dynamic behavior of wind generation as it relates to power system control and dynamic performance. Working group C4.601
29. Kundur P (1993) Power system stability and control. McGraw Hill, Inc, New York

30. Lalor G, Mullane A, O'Malley M (2005) Frequency control and wind turbine technologies. IEEE Trans Power Sys 20(4):1905–1913
31. Morren J, Haan SWH, Kling WL, Ferreira JA (2006) Wind turbines emulating inertia and supporting primary frequency control. IEEE Trans Power Sys 21(1):433–434
32. Ramtharan G, Ekanayake JB, Jenkins N (2007) Frequency support from doubly fed induction generator wind turbines. IET Renew Power Gen 1(1):3–9
33. Miller N, Clark K, Delmerico R, Cardinal M (2009) Windinertia[TM]: inertial response option for GE wind turbine generators. Presented at the 2009 IEEE power engineering society general meeting
34. Gautam D, Goel L, Ayyanar R, Vittal V, Harbour T (2011) Control strategy to mitigate the impact of reduced inertia due to doubly fed induction generators on large power systems. IEEE Trans Power Sys 26(1):214–224

Index

A
Ac–Dc, 22, 23, 30, 39, 39, 46
Active crow bar circuits, 60
Active power control, 67, 76, 90, 106
Active power control module, 108
Active power relationships in DFIG, 69
Aerodynamic model, 104
Air gap power, 68
Analysis of GSC in the synchronous frame, 90
Asynchronous injections, 125
Average model, 32–40
Average model implementation, 40, 49
Average pole output, 28

B
Beneficial impact, 122, 124, 126
Bi-positional switch, 53

C
Capacity of generation sources, 2
Carrier-based PWM, 52, 53, 59
CCA, 25, 26, 29–31, 33, 35, 37, 49, 50, 52
  implementation, 32
Component of wind turbine generator, 103
Components of VSC, 19
Configuration of DFIG, 66
Control of wind generators, 65
Constant speed operation, 103
Control
  block diagram, 81, 89
  functions, 58
  models, 110
Controller design, 21, 58, 65, 86

Controller design for rotor current loops, 88
Controller implementation, 21
Controls that are represented, 103
Converter gain, 42
Converter ratings of DFIG, 58
Converter topology, 22, 47, 50, 56
Crow bar circuits, 60, 62
Current
  command, 105, 138
  control, 87
  ripple, 37, 39
  stiff port, 34
Cycle-by-cycle average, 25, 28, 32, 91

D
Damping ratio, 117, 118, 121, 124, 126
DC link, 23, 29
  current, 29
  control, 93
  voltage, 93
DC motor drive, 34, 39
DC–AC, 22, 29, 39, 46
DC–DC, 19, 22, 29, 31, 33, 34
Decoupled control, 65, 86, 91
Detrimental impact, 123, 126
Development of Per-Phase equivalent
  circuit, 67
DFIG, 3, 6, 8–10, 12, 15, 16, 57–67, 69, 71,
  73–75, 78, 79, 81, 85, 86, 93, 94, 110,
  115–117, 119, 120, 122, 123, 125, 126,
  128, 130–138, 140, 141
DFIG as a Generator, 9, 10
DFIG control, 81, 115
DFIG equivalent circuit, 67

V. Vittal and R. Ayyanar, *Grid Integration and Dynamic Impact of Wind Energy*,
Power Electronics and Power Systems, DOI: 10.1007/978-1-4419-9323-6,
© Springer Science+Business Media New York 2013